AF816821

Digital Creator

Monetize Your Digital Assets

Myra E. Looring

Table of Contents

Introduction

Are you tired of working a monotonous 9–5 job every day? Do you see yourself drowning in debt, and struggling to make ends meet? Do you feel like you have no time for your personal life outside of work? If this is you, then this is the perfect book for you! You may be familiar with a few common ways you could make a passive income, but do you know there are many options out there that are viable and accessible to you?

We're living in a world where success and the way to achieve it have been revolutionized for the better. Anyone is capable of succeeding as their own boss with minimal capital, resources, and connections. It can all be achieved in the comfort of your own home on your reliable laptop. You're able to become your own boss, determine your working hours, and earn the amount of money you desire, but it all starts with the decisions you make!

Determine Your Future

"Take a risk and keep testing, because what works today won't work tomorrow, but what worked yesterday may work again."—Amrita Sahasrabudhe

At the end of the day, we're all in control of our own futures and the success we desire. This is why it's important to take the necessary risks to make a better future for yourself. Believe it or not, we're all able to create a tomorrow that works for us. It's all about finding what works.

Your success and financial security don't rely on what happened in your past or what is going to happen in the future. It depends on what you do in the present moment. It's important to live each day with thought and intent, so that you can work toward a future in which you're content.

What if you were told that you could quit that 9–5 job that's holding you back and pursue a career that provides you with multiple streams of passive income? This sounds like a dream to many people, but it can become your reality if you utilize the following tips in this book that'll help you to become a digital entrepreneur.

Out With the Old, in With the New

It's amazing to think about how we've evolved over the decades. I remember leaving university and my mom was so worried about me finding a good, secure job. All she wanted for me was to find a company where I could start at the bottom and work my way up to a better job, and sadly, just like my mom, this has been the mindset of many people. Work the same 9–5 job every day, hope to be promoted, and retire at the same company you've worked at your whole life!

When I graduated from university, I felt very lost. Yes, I had my business degree, but I didn't know in what direction it would take me. I was battling to find my path, as I knew I wanted to become an entrepreneur, but I was unsure of how I could make that happen with a lack of resources and finances. I felt as though I had no other option but to settle for a taxing 9–5 job that had nothing to do with my passion for business. Eventually, I got tired of this lifestyle, and knew that I had to find an easier, more enjoyable way to make money. This is when I came across the power of online businesses.

I saw how so many successful entrepreneurs were independently making money online, and I had to try it out for myself. I discovered that the world of business online is the way of the future, and from this I found out what tools I needed to start my journey toward digital success.

I started my online business over five years ago, and little did I know that this would be the beginning of my journey to success as an online entrepreneur. I'm now at a place where my online business works for me and provides me with a reliable stream of passive income. I've found a passion in selling digital assets, as it is affordable, simple, and rewarding.

This book will help you to realize your worth as a hard worker, as you'll no longer have to bust your butt for a boss who doesn't see your value and pays you less than you deserve. I'm thrilled to share my passion for digital monetization with you, as it's these tips and tricks that helped me to break away from my draining 9–5 job.

Life as we know it has been transformed, as it has become easier and more accessible for anyone to earn a six-figure income. This can all be achieved through having an online business and monetizing your digital assets. The beauty of becoming an online entrepreneur is that you are your own boss, and there are no expensive start-up fees, and you don't have to worry about any warehouses and suppliers. Although this work can be achieved by anyone, it still requires a lot of hard work and determination.

In this book, you will learn eight powerful online tools that can make you a successful digital entrepreneur. You'll learn what these tools are, and how you can use them to your advantage to maximize your profits. Through reading this informative book, you'll gain the knowledge and courage you need to leave that job holding you back, as you are guided onto a pursuit of success and freedom.

Chapter 1:

Merch by Amazon

Have you ever wanted to release your own merchandise, but you've never had the means to turn this dream into a reality? If so, then dropping merch with Amazon is the perfect way for you to make money off your great merchandise ideas. Merch by Amazon is an affordable and simple way to make a passive income. You're able to design custom merchandise, sell it, while staying within an affordable budget!

How to Get Started with Merch by Amazon

When you think of the concept of selling merch through Amazon, you may be deterred by how daunting it sounds. Maybe you don't have a lot of start-up money to put into your business idea, or you don't know how you'll be able to produce your merchandise. However, selling merch by Amazon is a lot simpler than you may think. This is a step-by-step guide that will help you to understand how to get started with merch by Amazon.

As with any business venture, the beginning can be the most challenging part. You have to learn how merch by Amazon works, and discover ways you can produce your merchandise successfully. You can follow these steps when you're getting started to ensure you're on the right track (Connolly, 2021).

Step 1: Create a Merch by Amazon Account

In order to sell your products through Amazon, you need to create an account with them. This is the foundation of your relationship with Amazon, as they learn everything they need to know about you. You need to apply for approval before they give you an account. You'll find this application when you visit merch.amazon.com: This application page will ask you for the following details:

- Personal details such as your name, age, etc.
- The type of industry for which you're making merchandise.
- Your organization or brand's name.
- The website that represents your brand.

Once you complete your application form, be prepared to wait a short period for a response from Amazon. You'll typically get a response within a week; however, it can be a bit longer, due to high demand from people wanting to sell through Amazon. Once you get your approval, you'll get your account, and you'll be able to get started.

Step 2: Create Your Product

When you've made your account, you'll be able to create your merchandise. It's a quick and pain-free process that requires you to do these following things to make your products:

- **Choose your form of product.** This means that you should choose the type of product you want to sell. For example, you can choose t-shirts, sweatshirts, phone cases, and pillow cases. Ensure that the product you choose is something you think your customers will like.
- **Create a design.** The design is the most important part of your product, as this is what differentiates your product from the competition. Make sure that the design you create is something unique that strongly depicts your brand. If you have a brand logo, you could use this as the design for your merchandise.
- **Upload it onto Amazon Merch.** Once you've created or found the perfect design for your product, it's time for you to upload it to Amazon Merch. This is when you need to make sure that your design meets all of Amazon's

requirements: It is no more than 25MB, uses RGB, and is a high-quality image.

Step 3: The Selling Stage

Once you've found both your product of choice and design and you've uploaded it onto Amazon Merch, you're ready to go live! This is when Amazon reviews your post and determines whether it's appropriate or not. These are the following things that Amazon will look for and do before making your product officially live:

- **Under review.** After submitting your design, it will show that it's going under review. If you haven't followed Amazon's Merch's guidelines, then your design is at risk of being rejected. This means that you'll have to find your mistakes, fix them, and resubmit a design that doesn't infringe on Amazon's content policy.
- **Processing.** Once your design is approved, it goes through processing. Your product will be added to their library, and they'll work on your listings. This is the final stage to expect before your product goes live.
- **Live.** If your design has gone through all of these steps, it will go live. This means that it's open to the public, and anyone can purchase your merchandise. However, if your design is

found to infringe on Amazon's policy, or if you don't sell a product within 180 days of it being live, then your design will be rejected and removed from their site.

Remember that the more you go through the process of making products through Amazon's Merch, the easier this will become to you. You'll find that getting your designs to go live will be a quick-and-easy process. You'll also become more successful as time goes, because you'll learn about what works and what doesn't work and eventually find your own unique brand that appeals to your customers.

How Does Merch by Amazon Work?

You now know how you can get started as an individual on Merch by Amazon, but now you may be wondering how this avenue of income will be profitable for both you and Amazon. It may seem too good to be true, as there is no risk of losing money, and you're able to fully profit from it. However, Merch by Amazon is just as straightforward and profitable as it sounds.

When you submit your design, and it goes live once it's approved, customers can view your products. Once your designs sell, Amazon will pay you royalties for your sale. All you have to do is choose a design, upload it, and set a price, and then Amazon will sort everything else for you. They will create the product listing, and when your design sells, they'll print it, package it, and ship it for you!

For some people making a design may be challenging, as well as you may not have the money to hire a graphic designer. This makes it a good idea for you to purchase your designs from businesses like Merch Ready Designs, and upload them directly to Merch by Amazon. It's as simple and affordable as that (Dunne, 2017)!

Tips for Success

Although starting with Merch by Amazon is simple and straightforward, there are still tips you can follow to ensure success. There are ways to ensure that you succeed on Amazon's Merch by making as many sales as possible. These are a few things to consider, so that you can succeed in this form of digital monetization.

Finding a Unique Brand

Having a strong brand is the most important thing to do when you want to succeed in selling merchandise. Because of how accessible Amazon Merch is to everyone, there is a lot of competition out there. This makes it crucial for you to find a unique brand that distinguishes you from your competition, so you appeal to a wider audience. These are some ways you can work on presenting a strong, unique brand on Merch by Amazon:

- **Understand your audience.** Your audience are your customers, so it's important for you to learn and understand your target audience. For example, if you're making products that are related to fitness, then you need to ensure that your products are targeted to people into fitness. For example, you could make designs with motivating quotes that could be put on gym t-shirts.
- **Have target niches.** The most important thing to remember when you're building a brand on Amazon's Merch, is that you need to use target niches. This means that you must stick to the niche you have chosen instead of using many and confusing your brand voice. It's also important to find a niche that you're passionate about. You could have a passion for being motivating, and this will cause you to make a line of products with motivational quotes. When you choose a niche you're passionate about, you're more likely to put your all into it. If you're battling to find phrases or designs to put onto your products, then you should visit

Pinterest and search for your niche, so that you can find inspiration.

- **Stay true to your brand.** It's important to have space to grow and evolve, but it's also important to always stay true to the essence of your brand. When you stick to your unique brand voice, it becomes stronger, and more appealing to customers, because they can truly resonate with it. When customers understand your brand voice, they're more likely to build a relationship with it, causing them to purchase your products.

Hire a Graphic Designer

It's always beneficial to get a professional helping hand when you're starting off with a new project. Sometimes, it's that extra touch that we need to create a stronger product the customers will love to buy. Hiring a graphic designer to make your design for your products is a great idea, as they'll be able to translate what you envision for your brand.

Although Merch by Amazon is appealing due to its low costs, it doesn't mean you should not spend money to make the best products possible! Having your design made by a professional graphic designer can give you an advantage over your competition. It's important to put your time, effort, and resources into making the best design, as it's your design that truly sells your product.

However, if you want to keep the budget low by not getting help from a graphic designer, you can find ways to make the design yourself. There are many applications online that will help you to create the perfect design at low costs. You could also make your design text-based, this way you can do it all on your own without any professional help.

Utilize Social Media

Instead of fearing the use of social media, use it to your advantage. Social media can help you in various ways to make products that are successful on Merch by Amazon. Although Amazon's Merch does all the work of selling your products on their platform, you should still create a presence online that will help you to get your products sold.

You should use a social media platform where most of your target audience would be active. For example, if you're an Instagram influencer wanting to sell your products to your followers, then you should be active on Instagram and promote your product via posts and stories on Instagram. By advertising your product and making an account for your brand on social media, you'll be able to reach a larger audience, and this can result in an increase of sales. Having this platform can allow you to alert your customers every time you release a new design.

Using social media is also important to find out the latest trends. When you're active on social media, you're able to keep up with these trends and find out what your customers want. You can also discover what works and what doesn't work by interacting with your followers. For example, if you're deciding between which design you should print on your new shirts, you could ask your followers which one they prefer. This helps you to release products that your customers actually want.

In this day and age, it's crucial to use social media when you're creating a brand, because this is how you reach the public, as well as a great way for you to determine what's in demand. If you're struggling to come up with concepts for your design, you can always check your competitors' socials for inspiration. The benefits of using social media as a brand are endless.

Why You Should Use Merch by Amazon

There are a number of ways in which you could make money through the sales of merchandise. You may be closed off to this avenue of passive income, because it doesn't seem like it's down your alley. However, you should consider all the following reasons why using Amazon's Merch would be useful for you. These are just a few ways you may consider when using Merch by Amazon.

You're a Public Figure

If you're a public figure of any kind, it's a good idea to monetize this asset. Being a public figure means that you have fans who support and love you, so you could always use this to your advantage by providing amazing products to your fanbase. As a public figure, you are your own brand, so it's important to release products that reflect this brand you represent online.

For example, you may be a well-known make-up artist on social media, and your content is filled with makeup videos, and tutorials. If you're looking to sell merchandise, then it's a good idea to make it related to makeup. You could release a hoodie that has a picture of you with your most iconic makeup look, or you could sell shirts that are artsy that portray your art skills.

You don't need millions of followers to become a public figure that people want to buy from. All you need to do is build a strong relationship with your audience online, and this will make them happy to support you. Remember that using a variety of different social media platforms will help you to get yourself out there more.

Expanding Your Business

You may own a well-established business, or you may be working on starting your dream business. Either way Amazon's Merch can aid the growth of your business. Having merchandise for your business is a great way of getting your brand out there, while making a profit from it at the same time.

If you have an established business with a loyal customer base, then selling merchandise can be very successful. There are many businesses out there that choose to sell merchandise to their customers, as it makes them feel as though they're a part of the company. It's also a great way for customers to show their support, especially if your business sells products that they want but the prices are high.

Selling merchandise when you're starting off your business is a great way to help establish your brand. It can help you to make a name for your business by getting it out there through merchandise. You can hold an event that promotes your new business, and you can give away your merchandise to people as a way to get them familiar with your brand.

People love merchandise, especially when it has a cool design or logo. It may be a great way to expand your business, and enhance your brand image. When you have clothing with your brand on it, it can make you appear more professional and established. This helps customers trust you and support your business.

Exploring a Passion

You may not have a brand or a business to use when selling merchandise through Amazon, but this doesn't mean that Merch by Amazon is not for you! You can use Amazon's Merch if you feel creative and inspired to make products of your own. Having your own line of products that you sell is a fun form of passive income, but it can be extremely costly and time consuming when you do it on your own. Amazon's Merch provides you with a platform that is affordable, easy for you to use, and pain free!

Why not take advantage of this opportunity if releasing merchandise is something you've always been interested in. It can help you to explore passionate aspects of yourself that you've never acknowledged before. You get to be creative when you come up with designs, and you get to explore a niche that excites you. For example, if you're passionate about health and fitness, you can use this digital asset as a way to express your passion with creativity.

Jensen's Success Story

You may be skeptical of using Merch by Amazon because it may sound too good to be true. Unfortunately, in the world we live in, when things sound too good to be true, they often are. However,

Merch by Amazon is really that affordable and simple! The story of Jensen's success with Merch by Amazon is proof that you can be very successful in this avenue of passive income.

Kim Jensen is a female entrepreneur who launched her business on Merch by Amazon. When she got started, she had no idea about designs and how to make demanded designs that people would like. She asked her daughter, who was studying design in college, to do her designs for her. This helped her to successfully make designs that caught customers' eyes.

Jensen has succeeded with her business on Merch by Amazon, as she has stayed dedicated. She achieved this by keeping up with the trends, and making designs that would follow them. Other tips that she gave that helped her succeed are as follows (Pilon, 2018):

- **Launch your designs early.** Not only did she keep up with trends and events happening around the world, she also made sure that she was one of the first businesses to produce these designs. By launching these trendy designs early, she received more sales than other businesses following the same trend. This is why it's important to monitor trends and stay on top of events, so that you can be the first in line!
- **Expand your business.** Jensen used Merch by Amazon as a stepping stone for her business, as this was how she began her journey as an

entrepreneur. However, she found success in her business by branching out to other avenues. Jensen has expanded her business by selling her designs through different platforms to ensure maximum success. She also used the revenue she made through Amazon's Merch to further grow her business in other ways. For example, she used it to advertise her Merch through Amazon or other platforms.

- **Start as soon as possible.** It's easy to get into the mindset where you tell yourself that you'll start tomorrow. The idea of starting with Merch by Amazon intrigues you, but you're scared to commit to it. The leading factor that made Jenson succeed in Amazon's Merch is the fact that she started immediately. She didn't second guess whether she should do it; instead, she threw herself into this project. The benefit of Merch by Amazon is that you can start as soon as possible because you don't require a lot of money to start up.

However, Jensen's journey on Merch by Amazon wasn't all that easy, as it took time and dedication. When she started off, she struggled, as she wasn't making the sales she wanted. She was unaware of what the demand was and how to appeal to a target audience. But with time and dedication, she learned the ins and outs of selling through Merch by Amazon. It's because of her perseverance and hard work that she managed to list new products often that she knew were in high demand.

Chapter 2:

Etsy

We've established that you can sell products online through Merch by Amazon, but what if you were told that you could sell something that people can simply download! Etsy is a platform that allows sellers to sell digital assets that are in the form of a download. The customers on this platform look for a variety of unique goods.

What Is Etsy?

Etsy is a platform that connects independent sellers with people looking to buy a variety of different digital goods. You can find almost anything you're looking for on Etsy, meaning that there's little to no limits on what you could sell through Etsy. Much like Merch by Amazon, you don't handle anything in the process of you selling these digital goods.

How to Sell on Etsy

Etsy sellers are independent individuals who sell their unique projects and works of art. It's simple and quick to get started on Etsy, as all you need to do is create an account with Etsy. From there, you upload your project onto its platform, and you're charged a small listing fee of $0.20 for four months. You can sell a large variety of goods, whether they're physical products or something customers can simply download.

Once someone purchases your item, you get paid for it. Etsy takes a 5% transaction fee (price + shipping) and a 3% + $0.25 payment fee. Then you can retrieve your money through a number of payment methods such as PayPal, debit or credit cards, an Etsy voucher, etc. Once this is complete, you've successfully sold something through Etsy.

The benefits of using Etsy are that it's easy to get started and is an affordable way for you to make sales. Also, when shipping, the purchased item goes straight from you to the customer.

As a consumer, you should consider buying your artistic goods through Etsy, because you'll be supporting independent sellers. These sellers put a lot of time and effort into their unique goods, and many of them are of a higher quality than similar products you can buy from big companies.

Advertising

The process of adding a listing to your Etsy account is quite simple; however, it may be a bit challenging to successfully sell your items, especially when starting off. As with any business, it's important to advertise to get your products across to a wider audience. When you're an individual artist selling through Etsy, remember that selling your craft is your business, so it may be worthwhile to spend money on advertising.

Etsy provides advertising services that promote your products on its platform, so that you can reach more potential customers. They also offer off-site ads, where your product will be advertised to people who are searching for your product on the Internet. However, there is a fee for this advertisement, but it's a worthwhile investment.

Spending that extra money to get your product advertised will attract more action on your listing, making you more likely to sell your products successfully. Not only should you pay for advertising, but you should also work on creating an outstanding brand that attracts customers. After you work on your brand, create a shop that is eye-catching, build your following, and use Etsy's advertising services. Then you'll be on the right track toward success.

The Best Digital Products to Sell Through Etsy

Etsy can be used to sell a variety of products. Many creatives choose it as their way to independently sell their artworks, crafts, and antique items. This may seem like it's something out of your expertise; however, there are still ways for you to make money through Etsy if you're not physically artistic. What makes Etsy so special is that you can sell downloads of projects, instead of physical products. There's a long list of downloadable goods that you could sell with ease (Mama & Money, 2020).

Benefits of Selling Goods in Digital Form

In order for you to be successful in selling these forms of digital assets, you'll need to discover what's in demand and how you can create it. This doesn't necessarily take skill or talent to create, but it takes time to create the best digital assets worth purchasing. You can only benefit from venturing into selling these types of digital assets. These are the following benefits of selling goods in a digital form:

- **Provides passive income.** Unlike the unique, individual products you can sell through Etsy, digital products are something you only need to work on once. For example, you may create artwork in a digital form that can be used as wallpaper. You only spend time on making this once, but it can be sold numerous times. This is a form of passive income that you don't have to pay attention to, as you'll make money from your digital assets without doing anything.

- **Highly profitable.** Due to this being a form of passive income, it is completely profitable, meaning that most or even all of the money you get from it is your profit. This is because you can spend little to no money on creating your digital asset, making your revenue pure profit. This is a great side hustle because you don't need a lot of money to put into it, while you still make a substantial profit from it.
- **Large variety.** We all have different interests and skills, which can make it difficult to find a source of income that everyone can use. This is what makes Etsy so useful to everyone, as there is a large variety of digital assets you can sell. You can even choose to sell different types of digital assets, so that you can ensure a better success rate. Having that variety means that you're bound to find a digital asset to make that you'll enjoy. However, it's also important to sell digital assets that are in demand, so that you know it will get sold. These are just a few digital assets that have a high success rate.

Art Printables

Art printables are one of the best digital assets to sell, as they are in high demand and easy to make. You can sell pretty, well-designed art printables, or you could sell a printable with an inspiring quote and a cool background—it's as simple as that!

If you are a creative individual, you can express this through your art printables by making fun pieces of art. This can be very popular as people may use it as their wallpaper, send it to friends or family, or set it as their display pictures on their social media platforms.

Even if you don't have any artistic ability, you can still create these art printables. You could either collaborate with an artist you know, or you could make motivational printables, as they are quite popular. These printables can have inspiring quotes that you can get from Canva.

Checklists

If you're a well-organized individual, who loves to make lists and plan events, this may be the perfect digital asset for you. Believe it or not, there is a high demand for checklists out there. People are willing to pay money to get a checklist that helps them to plan their events efficiently.

If someone is getting married, they may be overwhelmed by all the different things they have to think about and do. This can be overwhelming, especially because it's an exciting day you want to celebrate (not stress about). This can make purchasing a general checklist of what you should do extremely helpful. Having a tangible list of things to do can help them feel more organized and in control of their wedding.

Weddings aren't the only checklists that are in demand. You could fill your Etsy account with different types of checklists people may want. You could have a checklist for birthdays, anniversaries, or any form of party.

If you like the idea of helping people plan their events, but you don't want the pressure of making their to-do checklist, then you could always make spreadsheets that they can use. Making a well-structured, easy-to-use spreadsheet can help your customers to plan their own events with ease!

Invitations or Cards

We're living in a day and age where physical copies of invitations and cards are no longer given out. This means that there's a higher demand for digital forms of these items. People can save money on delivering their invitations and cards to each other, and they can purchase a digital copy that they can send via a text message.

You could create invitation templates to parties or weddings, and this gives the customer the freedom to address them to their visitor, and write whatever they please. You could also offer a large range of cards that people can customize.

This is a simple digital asset to sell that will always be in demand. You can create these cards and invitations by using affordable software online. Remember that spending a small amount of money making the digital asset is worthwhile, because once you put it on Etsy, you could make a sizable profit.

Tips for Success

Etsy is an easy-to-understand platform that will help you on your journey toward opening your Etsy shop. They'll direct you in your first listing, and all the directions are clear. However, there are some things to consider and remember if you want to succeed on this platform. These are some things you should consider when you're starting your Etsy shop:

- **Be unique.** Etsy is a platform with many independent sellers, meaning that you're bound to compete with someone. This means that your products and brand need to stand out, so that customers are drawn to you. Being unique and different can be difficult, but it's simpler when you find ways to express yourself through your products. **Build your audience.** You may have an Etsy shop where you sell a variety of digital assets that reach a wide audience, but it's important to ensure that you create a brand that helps you to build a strong audience. This can be achieved by being authentic and genuine to your customers and providing them with products they want.

- **Upload files efficiently and safely.** You should use zipping files and reduce your image as much as possible, so that your file can be relatively small to download. You may be using Adobe Illustrator to make your digital product, so it's important for you to turn off the editing option on your PDF. This is important as it'll make your file smaller and prevent people from stealing your content.
- **Do your research.** Above all, it's important to do your market research! You need to determine whether a product you're going to list on your shop is seasonal, or evergreen. Seasonal means that it will sell well in a specific time in the year, as it follows a trend or event happening. In contrast, evergreen means that there will always be demand for the product. It's a good idea to sell seasonal goods as they can do well, but it's also vital for your shop to

have evergreen products that'll always be sought after.

What Can You Sell?

The above examples of digital assets are merely a few popular options you can explore. However, you don't have to focus only on the best sellers, as you can branch out to selling the many different types of digital assets there are. It's better to work on a project that excites you, because that way, you'll make something successful that customers want to purchase.

There are approximately 23,000 digital assets being sold on Etsy, costing around $8 on average. I've made a steady monthly income between $50—$200 without having to work on it every day. What's stopping you from starting your own Etsy shop? After you see the numerous ways you can make money through digital assets on Etsy, you're bound to want to start your own shop (*The ultimate list of digital products to sell on Etsy*, 2019)!

Cut Files

This is for people who release t-shirts and decor with designs on them. Essentially, you could make the designs for people who use Merch by Amazon. All you need to do is make a design that people will like. You could find a niche to stick to, so that you can build a strong customer base.

For example, you may target designs for home decor and the theme you choose for it may be love. This means that you should make designs that portray this theme, so that people releasing home decor with this theme purchase your designs. You could make simple designs such as a heart with the word 'Love' written in a cool font.

You could also make designs with quotes, as they are popular on Etsy. When you use quotes, you must check what's copyrighted, and avoid using those quotes. Use an interesting font to differentiate you from your competition, or you could even hand-write your quotes to make them more authentic.

Printable and Digital Planners

Not only can you make checklists and checklist templates for events, but you can also make planners that help people keep track of their everyday lives. We all have busy schedules, where it seems like we have a million and one things to do. Using a planner can be exactly what you need to gain back control over your life. This can make a planner a popular and useful digital product to sell.

The perfect planner is well organized, appealing to the eye, and has enough space to jot down plans. You generally want to make a planner that's blank, so that the customer can fill in their own plans; but this doesn't mean that your planner shouldn't have any personality. Making your planner vibrant and motivating can make it stand out to people looking for the best planner to purchase.

You can make your planner digital or printable—it depends on your audience. You may know that your customers are people who do everything on their phones. In this case, you'll want to make a digital planner that the customer can download and adjust on their phone or their laptop. You need to make your planner adjustable so that they can type in their own schedules.

If your customers are more old fashioned and want to see their schedule on paper in black in white, then you should make a printable planner. This means that your customer can purchase your planner, download it, and then print it. They can use the hardcopy instead of using a device. When you make a printable planner, it's important to leave enough space for the customer to write their plans for each day.

Coloring Book Pages

Coloring books have become highly demanded lately, and they're not just for children's use anymore. It's become a trend for adults to also use coloring books, as it's a therapeutic activity they use to escape their stressful lives. You can become successful on Etsy by selling coloring book pages.

You can sell adult coloring book pages that have different designs and patterns. Even though adult coloring books are trending currently, why limit yourself? You could also sell children's coloring book pages, and even create a package where you sell both an adult and child coloring page at a discounted price.

Buying coloring book pages through Etsy would be appealing to customers, because physical copies of adult coloring books are quite costly. It's more affordable buying the download, and printing it yourself, instead of buying a hardcopy at an arts and crafts store.

eBooks and Guides

If you're someone who enjoys writing, then selling eBooks or guides on Etsy may be for you! You may not see yourself as an award-winning author, but if you have a good standard of English, you'll be able to write interesting eBooks and guides that people would love to purchase!

People love books and guides that help them to learn how to do new things, and this can be a great way for you to teach people about things you're passionate about. There's a large variety of niches and topics you can write about, including the following:

- **How-to guides.** You can make a how-to guide about absolutely anything, and it could be useful to many people. A good way to find out what people want to learn how to do is by doing a google search and finding the top searched results. Understanding what people need guidance on out there will help you to make guides that are in high demand.
- **Recipe books.** You can write cookbooks with a variety of your favorite recipes! People love to find new inspiration when cooking, so make sure that your cookbook is filled with exciting recipes that are different and delicious.
- **Health and fitness guides.** If you're someone who's into health and fitness, then you can write guides that help people adapt to a healthy lifestyle. You can give them a workout routine that you used and add healthy meal plans they could follow. When you're making these types of guides, make sure that you're doing your research, as you don't want to share false information.

Website Design Templates

Knowing how to build a website is a valuable skill, and many people looking to start their own websites will pay for your help! You can create website design templates, so that customers can purchase them and directly upload them to a website they've created on platforms such as WordPress and Squarespace.

You don't have to be a tech wiz to learn how to design a good website design template, as it can be very straightforward. These are some tips that will help you to learn how you can successfully build a web design that your customers can purchase (Abenzour, 2019):

- **Understanding your customers.** When you're considering building website design templates, you need to take the time to get to know and understand the market. Who is going to buy your product, and what is the demand out there? Knowing your customers will help you to sell your website design templates successfully. You may have customers who are freelancers and want to build portfolios for their photography, writing, or etc. Understanding their wants can help you to make the best templates that are in demand.
- **Choose an appropriate template style.** How you present your template is the most vital aspect to a website design. Take time to consider the different fonts, images, and page layouts you use, as they all contribute to the aesthetics of the website design. Something as small as the color on your page background can affect who's interested in your product. It's

important to make your web designs look professional, yet fun and eye-catching. This will help you to stand out from other website design templates.

- **Make it mobile friendly.** It's more common for people searching on the Internet to be using their phones, rather than a desktop. This is why it's crucial for you to make your website template mobile friendly. Ensure that the text, icons, and images can all be seen properly on mobiles as well.

Meal Planning Templates

We've established that people like to purchase templates that can help them to plan their lives. Another popular template to sell is a meal planning template, where people can write down different recipes and meal ideas that they can cook. A meal planning template can be designed in a number of ways.

You can design a weekly planner where you write down each day of the week and leave space for the customer to write in their meal plan for the day. This is perfect for both printable and digital planners, as the writer can reprint the meal planner each week, or they could retype their meal plan.

Your meal plans can also be designed in a more casual way. You can set the planner out for it to be perfect to write down new recipes and meal ideas. Customers can print this out as many times as they please, and they could end up with their own cookbook filled with recipes they wrote down on your meal plan design.

This digital asset is a good idea, not only because it'll help people to stay organized and planned, but it can also promote a healthy lifestyle. If people have a meal planner to help plan their meals, they're more likely to stick to a healthy diet.

Chapter 3:

Fiverr and Upwork

Did you know that you could have a successful side hustle as a freelancer? Many successful entrepreneurs gain another source of income through their freelance work. Becoming a freelancer is a lot easier than you may think, as all you need to do is find the right sites that will help you start your journey. Reliable platforms that connect freelancers with people looking for work done independently are Fiverr and Upwork.

How Fiverr Works

Fiverr is a platform that connects freelancers with people looking for independent, more affordable services. These buyers pay for freelancer's gigs in advance, and then the freelancer delivers. The freelancer gets paid 80% of what the buyer pays, and the remaining 20% is the fee for Fiverr's services.

Fiverr is such a useful platform to get started with as a freelancer, because it provides you with a platform where you can work directly with customers without having to worry about the hiring and firing process of jobs! All you have to do to get started with Fiverr is start an account with them, and you're set to go. However, there are some things you should consider when getting started, so that you can ensure a successful freelancing career.

Becoming a Successful Seller

Did you know that you could earn approximately $2,000 a month just from freelancing through Fiverr? You're able to make a living off Fiverr as there's an endless number of buyers looking for freelancers to work on their gigs. You can also set the standard for how much you want to be paid, as you are in control of setting prices for your gigs. All these elements make it easy to succeed through Fiverr; however, there are some tips you should follow to increase your chances of becoming a consistently successful freelancer (Schulz, 2021):

- **Build your portfolio.** As a freelancer, you are your own business, and this is why it's vital for you to market yourself accordingly. You can achieve this by building your portfolio, i.e., your seller ratings. Seller ratings are reviews and score sellers get for the delivery of their gigs. The more projects you do, the more seller ratings you get, and therefore the more accredited you are on Fiverr. Your ratings are equivalent to your resume as it's what buyers look for to check if you're experienced. The better your portfolio is, the more gigs you will sell, and you'll be able to raise the prices of your services.

- **Offer add-ons.** Less is more, but sometimes more is better! If you offer add-ons to your services on your gigs, you can increase your asking price. For example, if your freelancing job is photography and a client wants you to provide them with stock images, then you could add on detailed descriptions of your photos. Offering this adds on to your clients and makes you sound more professional, and it'll bring in extra money.
- **Use AND.CO.** This software helps you to stay on track with the ins and outs of your freelancing career. This is an important asset for any freelancer to have, especially if you're a beginner. It helps you to keep track of your invoices, payments, proposals, and everything else that you need to organize as a freelancer. This is extremely helpful, as it takes away the burden of remembering everything that you need to do, and you can focus your time and energy on your gigs. You can download the free version of software, or you can upgrade to the paying option with more helpful additions.

Best Ways to Make Money Through Fiverr

You may be deterred from using Fiverr because you don't think you have any talents or skills to use to become a freelancer. What makes Fiverr so special is that it's so versatile, as it has a long list of the different freelancers they need. This is due to a large number of businesses seeking services from freelancers, rather than business, as it's more affordable.

After viewing the long list of freelance roles out there, you're bound to find something that suits you. You may even find something that you've always been interested in trying—now's your chance! The beauty of Fiverr is that you can start as a freelancer on any level. You could be a beginner starting off your career, or you could be an expert, looking for a more casual, free way to work. Regardless of your expertise level, you'll find gigs that are perfect for you and your freelancing career.

It's important to find out all the different freelancing roles out there, as there are over 300 digital services on Fiverr. These are merely a few freelancing jobs that are in high demand that you could try out. Do your research to discover the perfect freelance role that best suits you (Dean, 2019).

Research

Do you enjoy learning new information? Are you someone who is constantly searching topics online for an answer? If this is you, then you may be interested in selling your research skills. You could get paid for working on research projects through Fiverr, and you just have to be a fast researcher who can turn in research projects at a fast rate.

If you enjoy research, but you don't know whether you have what it takes to be a freelance researcher, then you can always work on developing your skills. These are some ways you can become an efficient researcher:

- **Developing your topic.** You'll be given a topic by your client, and you need to work on developing it before you dig into your research. This can be achieved by breaking down your topic. Start by finding the keywords in your topic, and ensure that you understand the definitions of all these important words. After expanding your topic and making sense of it, it's time to write down important research questions that you'll answer throughout your researching process.
- **Use your sources.** The more sources you use while researching, the more valuable your research will be. It's important to branch out and use as many sources as possible to answer

your topic questions. When you use multiple sources for your research, you're able to fact check the different information you've found. It also helps you to have a large amount of research for your client, which is always better than having too little!

- **Organize your information.** You don't want to give your client research that is disorganized and all over the place. They will probably specify how they want their research set out, but regardless of their preference, they will want something written out in a formal, organized, and concise structure. You should do your research on a piece of paper, or a separate google document, if you choose to type out your rough research. Once you've finished doing all of your research, you should type it out, ensuring that you have headings, paragraphs, bullet points, and lists so that you can keep a good structure for your final product.

Virtual Assistant

This freelancing role is for all the organized individuals who can make well-planned schedules and stay on top of both their lives and those of their clients. As a virtual assistant, you need to be disciplined, so that you stay committed to your role. You also need to be good at communicating so that you can efficiently communicate with your client about their daily schedules. These are just a few tasks that would be expected from you as a virtual assistant:

- **Scheduling appointments.** As a virtual assistant, you're responsible for scheduling the meetings, activities, and appointments of your client's day. You need to be really organized when doing this, so that you don't overlap their appointments. You also want to schedule their day in a well-structured way, so that they still have the time to rest and eat food.
- **Making phone calls.** This is why it's important for you to have good communication skills as a virtual assistant, as you'll need to make all the calls for your client. They will let you know who they need to be called, and what for. They may even give your number out as their assistant, so you'll receive calls on their behalf. When on the phone on behalf of your client, you need to be professional and friendly.
- **Managing emails.** You need to be a trustworthy individual, as you may get access to your client's important work emails. You'll have the duty of answering their emails, alerting them of anything important, and clearing out any unwanted junk mail.

The benefits of being a virtual assistant is that you get paid well, and you don't even have to leave your home! Everything you need is on your laptop and phone, so you're able to manage your client's life from any location.

Write Reviews

Reviews are an important part of any business, because they show potential customers the quality of their products or services. This is why review writing is a highly demanded job on Fiverr. You could get the opportunity to review a number of different products and services.

Anyone can write reviews. All you need is a good standard of English and be able to write at a fast rate. If you're able to produce 400 words in approximately 12 minutes, then this is the perfect freelancing job for you!

Social Media Management

Being able to manage social media accounts is an important skill to have when you're running a business; however, not every business owner possesses the skill or time to do it correctly. This is where your social media skills will come in handy. You can do the following things to help grow and build your client's social media accounts:

- **Understanding engagement.** When you are aiming to build your social media accounts you need to understand your engagement. Doing some research and investigating your client's accounts can help you to determine their audience, their peak times of engagement, and other important factors that can impact your growth.
- **Consistent posting.** In order to grow your client's following, you need to ensure that they're always active on their accounts. You can give them ideas of content that they could post, or you could work on creating the content yourself (depending on what your client wants). Once you have a pattern of posting, at peak engagement hours, you'll notice the followers are increasing.
- **Writing captions.** Catchy, vibrant, and friendly captions can make a huge difference to a business's socials. If their captions show a sense of humor and friendliness, people will be more likely to connect with the brand. This will grow your client's socials and could result in an increase of paying customers.

Top 5 Lucrative Freelancing Jobs on Fiverr

We've covered a few freelancing jobs that anyone can get to, but are you a high achiever who wants to make the most of this platform? If you're like me and you want to start earning a steady $2,000 a month or more, then you can look into these top 5 paying freelancing jobs on Fiverr:

1. **Video marketing.** Video marketing is making or posting videos that promote a product or service. Freelancers on Fiverr make anything from $3,000–$18,000 per project.
2. **Mobile app development.** The next best lucrative side hustle is mobile app developing, and for this you need some experience with coding. Sellers often charge $300–$3,000 per app they develop.
3. **Website building.** The demand for websites has increased, as every business wants a website to represent their brand. Buyers are willing to pay from $100 to $3,000 for a built website.
4. **Graphic design.** If you're a skilled graphic designer, your talents are highly sought after. You can create designs for a number of different projects, and get paid between $100 and $3,000 per project.
5. **Search and display marketing.** This form of marketing requires you to understand how to

work with Google's advertising service online. You can consult your clients on how they can optimize their advertisements on Google, or you can physically do it for them. This service is sold for $100–$2,000.

These freelancing jobs take more skill and effort, but that doesn't mean you're not capable of exploring them. You could choose one that is similar to work you've done in the past, or you could pick something completely out of your comfort zone and take courses. There's nothing you can't learn with a helpful course that teaches you the skills you require.

How Upwork Works

Upwork is another platform that connects freelancers with companies that are seeking freelancers. As a freelancer on Upwork, you're able to work on as many projects as you like, as you set the pace of how much work you do. If you have the time to work every day in a week, you could take on more projects, but if you're busy with your main job during the week, you could take on a few projects a month. It's all up to you and what you can handle.

Upwork is quite similar to Fiverr, as you set your own hourly rates, and you have a large variety of different projects to choose from. However, the process of landing the project is a little different. You can visit Upwork and proceed to fill in an application. It will take around 24 hours for Upwork to review your application and accept it. Once you're accepted, you will be able to view the long list of jobs that are available.

These are some categories that Upwork has to offer to their freelancers. You can consider any of these freelancing roles regardless of whether you're a beginner with little to no experience, or a well-skilled expert, as long as you have a passion and interest for your work (Helms, 2018).

Design and Creative

If you consider yourself a creative individual, why not turn this into your freelancing side hustle? You could be making money off an interest or skill that you've never utilized before. There are many different types of freelancing positions that involve design and creativity. These are a few things you may have interest in:

- **Music**. There is demand for freelancers in the music industry. If you're someone who has experience or passion for producing music and audios, you can consider becoming a freelancer in this category. You may be provided with a number of different job openings such as ghost-writing songs, working on beats, or producing songs.

- **Photography.** Getting into photography is fairly simple when you have the eye for it. All you need to do is invest your money into a good-quality camera, and then you can start your photography journey through freelancing. You may find work that you can provide digitally, or you may find jobs where they require you to physically be there. This could be the perfect opportunity to travel and work!
- **Illustrating and art.** There's high demand for creative artworks and illustrations. Clients may need art for a number of different projects, and paying for it from a freelancer would be more affordable for them. They may need artworks that they outsource and sell as digital assets, or they may need illustrations for an animation they're working on.
- **Voice talent.** Have you always been complimented on your voice? Do you have a soothing voice and are you able to do voice acting? If so, you may be considered as voice talent. The type of projects under voice talent are voice-over jobs for adverts, videos, and many more! You could also get projects where you record audio books.

Legal

Did you know that you could do freelancing work that involved legalities? If you have experience in law and have a lot of knowledge about various legal issues, you may want to consider taking it up as freelance work on Upwork.

Many companies and individuals seek legal help through Upwork, because it's more affordable than approaching a legal firm. All you need is some experience and a background in the legal industry. You may even be a full-time lawyer who wants to do rewarding freelance work on the side, as you're able to use your skills to impact many individual lives.

Marketing

Marketing is the way of the future, as you need to implement marketing campaigns whenever you're aiming to succeed in business. If you have experience with marketing, or you've studied marketing, then you could consider offering your freelance marketing services.

You may be experienced at marketing, as you've had to market yourself as a freelancer, or you've done a few courses on different types of marketing. These are the kind of jobs you may be expected to do as a freelance marketer on Upwork:

- **Digital marketing.** Nowadays everyone has a device they use for everything they do. This is why it's crucial for businesses to invest money in digital marketing. This form of marketing entails the form of advertising that happens online, whether it's an advert that appears on google, or in someone's emails, it's all a form of digital marketing.
- **Social media marketing.** It's important as a business to have a presence on social media. If you're a social media expert, then you can help businesses reach their social media goals. They may ask for help creating social media campaigns, building followers, strengthening engagement, and creating quality content for social media.
- **Content marketing.** Content marketing can entail putting together content for social media, websites, and blogs, etc. If you enjoy writing and creating content that can help to promote businesses, then you can consider doing freelancing in this field.

Writing

There are different forms of writing that are in demand on Upwork. You may not see yourself as a writer, but there could be a writing service that you're capable of doing. There will always be a need for freelance writers, as it's the main source of content out there. If writing is something that you enjoy, and you're relatively good at it, you should consider it as a freelancing job. However, it can be a time-consuming side hustle, so make sure it's something you're interested in.

These are some types of writing services you can expect to find on Upwork:

- **SEO writing.** SEO stands for Search Engine Optimization, and this means that articles are written in a certain way to optimize its views when people are searching for related content. This helps your article to stand out, and get more views, without having to pay for advertisement. Many businesses and blogs need SEO writers to help get their content out there and succeed in their field.
- **Scripts.** Scripts are some of the most enjoyable things to write, because it can be a fun and creative process. There are many different types of script writing jobs such as writing scripts for YouTube channels, scripts for video games, and many more!
- **eBooks**. You'll see a variety of eBook projects up for grabs, and you can choose to pick up one that you feel most comfortable writing. You could write non-fiction books with a variety of different niches such as health and fitness, self-

help, and business. If you enjoy more creative projects, then you can take on a fiction eBook project.

- **Factual writing.** If you're a good researcher who enjoys writing, then you can write factual articles. There are many projects on Upwork that require you to be knowledgeable or good at researching. You need to be able to write in a formal and informative tone, so that you create a factual article that people can believe and understand.
- **Resumes and cover letters.** You have written countless cover letters for your many job opportunities, and you always update and pay special attention to your resume. If this sounds like you, then you can pick up jobs on Upwork where you can write resumes and cover letters for people who need to sell themselves for their dream job applications.

Chapter 4:

PLR Method

If you're anything like me when I first started becoming a digital creator, you have zero clue as to what PLR content is. PLR stands for Private Label Rights, and it means that content that gives you the right to edit or change it how you please. This may sound unusual and unprofitable, but there are a lot of ways to make money through PLR content.

What Is PLR?

We live in a world filled with creativity and new ideas, but due to a large number of these ideas, it can make it almost impossible to come up with something that is completely different. This is what makes PLR content so helpful, because you can use already made content and adapt to make it more unique

Due to this form of creating content, you're able to make content quickly without having to go through a long creative brainstorm process. It's a time efficient way of creating content, without plagiarizing and copying someone else's work. The PLR method is applicable to numerous forms of content creations such as the following (Sheldon, n.d.):

Written content

There is a variety of written content out there that can help you grow your business, blog, and social media accounts. It's important to write engaging content that hooks in readers. If you're creating a website for your business, it would be a good investment to start a blog that you regularly post on the site. You can post blog articles that are related to your products and find ways to promote them through these posts.

Finding content to use to create these blog articles will help you to write professional articles that draw in customers. You'll find articles that successfully use SEO, which will help you to publish blog articles that reach a wider audience. Starting a business in this day and age requires you to be an active content creator, so it's important to use PLR to make this process easier and more successful for you.

You can also buy rights to written content such as emails and all types of articles and eBooks. It can be difficult to start writing any form of content from scratch. It requires a lot of brainstorming and creative thinking, but with the use of PLR, you'll be able to make this process more manageable for yourself, as you'll be able to produce content that is of high quality that achieves desirable results.

Videos

Although written content is a vital form of content that people will always need, video content is becoming increasingly popular and fundamental. In a world where we're all busy and occupied, being able to watch a video instead of spending time on reading is just what people need. Being able to buy the rights for videos allows you to make similar content that captivates audiences and gives your business, website, and social media accounts better engagement.

When you're creating a website, it's important to add videos so that it can increase your engagement. This helps readers of your article to have a visual perspective of what you're trying to say in your article. It's also a good tool that helps promote SEO marketing for your published article.

Audios

Another popular form of content that people tend to rely on, due to its time efficiency, is audio. If you're planning on making long forms of written content such as eBooks, then you should consider making audiobooks. Audiobooks are basically eBooks in audio form. This is useful, as many people out there want to read books, but they don't have the time. Having access to audio versions of these books helps them to read a book they're interested in, without spending the time or energy on physically reading it.

You can also buy rights to other forms of audios that you may need for advertisement purposes, podcasts, or video backgrounds. The possibilities are endless for audio creations. Once you buy rights to an audio you want, you can re-record the audio, and add your own elements to it.

Graphics

When you're starting your life off as an entrepreneur, you're going to need graphics for a number of reasons. If you're starting a business, you'll need a logo that represents your business. This can be difficult to create, especially if you lack experience in business. Buying the rights to graphics that have designs that intrigue you, and interpret exactly what you see for your business, helps you to use a logo that'll create a holistically sound brand.

You may also need graphics for your websites, eBook covers, blog backgrounds, and many other projects you're working on. Unfortunately, we don't all know a graphic designer who can make graphics for us, so it's helpful to buy rights to designs that match your brand's aesthetics.

Online Courses

Maybe you're interested in designing an online course on something you're passionate about, but you don't know how to go about making it informative and easy to follow. If this is something you're struggling with, you can buy the rights to this type of content and adapt it.

When making an online course, you don't need to buy rights to an online course that's exactly what you want to create, as this may be hard to find. The best thing to find is something similar that'll give you an accurate guideline when you're making your online course.

You may be interested in creating an online course that educates and helps others, but you don't know what to make that'll be successful and useful. Here are few online courses that you could buy rights to and make yourself:

- **Computer skills.** It's no surprise that computer literacy is a requirement for many jobs, as the use of technology in the work environment has become something none of us can avoid. This is why many people are willing to pay money to participate in online courses that teach them the necessary computer skills they need to excel in any environment.
- **Learning how to learn.** If you want to be successful in life, you need to accept that the

road to learning is never ending! This is why online courses about learning can be useful to a wide variety of people. It can be used by people who are learning a new industry, or it can be used by individuals who want to learn how to be better at what they do. These online courses are also valuable to students who need learning skills, so that they can pass their tests and assignments.

- **Psychology.** Many people have a fascination with the mind, but they're not willing to study for a full three-year degree at university. This makes a short online course attainable and attractive to them. There are many different types of psychology to venture into, which makes it an exciting and vast subject to teach. When creating this course, you must do lots of research, as well as ask for tips and help from a certified psychologist.
- **English skills.** If English is one of your strengths, then why not help others become as fluent and grammatical as you? You can create a course that helps people learn English if it's not their mother tongue, or you could help native English speakers develop their English skills. Some people want to improve their English so that they can write more, for a job, or for public speaking purposes.

A great way to learn how to make your own online courses is by attending some of your own. Content Sparks is a platform that helps people to reach their content goals online. They have a very useful online course that'll help you to get into PLR content. Taking this course will help you to learn how to make your own online courses, as well as it'll help you to learn more about PLR content.

Benefits of Using PLR Content

Buying the rights to other people's content is a great way to find some inspiration to make content, especially if you're battling to come up with new and fresh content. There's a long list of benefits that come from making this decision, such as the following (Thackston, 2020):

- **Less researching.** If you're like me, then you find endless hours of research tedious and boring. When you buy the rights to content, you save the trouble of having to do tons and tons of research. You may still have to do research to see if there's anything you'd like to add or remove from the content.
- **Engaging content.** By using PLR content, you're able to keep up to date with the trends and interests of your audience or readers. It can be difficult to know what people want to see, what videos they want to watch, what

campaigns they want to follow, and what articles they'll feel interested in reading. When you use PLR content you can use what you think will do well, and you can learn what works and what doesn't work through trial and error.

- **No more writer's block.** We've all experienced writer's block, especially when we have a new exciting, creative idea. This may be a once-in-a-lifetime idea, but your brain is finding it impossible to come up with ways to execute it. When using PLR content, you never have to worry about this creative block, as there may be the perfect creative form of content that will express exactly what you're looking for. If not, you can always adapt a piece of content into something more suitable for you.

Although there is a long list of pros that come with using PLR content, there are also cons, especially when you don't do it right. This is why it's important to use the right resources, so that you can successfully use PLR content to your advantage.

There are countless sources for PLR content that you can find, so it's important to do your research to find the best possible resources. As mentioned, Content Sparks is a great platform that can help you to reach your content goals. There are many different platforms and websites that will provide you with the opportunity to learn more about PLR content, as well as they'll provide you with the necessary content you're looking to buy rights to. This is how Content Sparks can help you on your journey:

- **You can use it for your clients or students.** If you're looking to make courses for either your clients or students, you can do it through a content site that provides PLR products. You can either be a teacher who releases a variety of courses to your students, or you could sell these courses to people who want to sell them to their own students.
- **Attend their courses to understand PLR content better.** PLR content can be tricky to understand at first. If you're considering getting started with PLR content, it may be worthwhile for you to take a short course that teaches you tips and tricks that'll make you more successful.
- **Provide you with useful materials.** Content Sparks provides you with PLR content that you can turn into your own. Not only do they provide you with courses that you can take, but they also give you templates for your own courses that you can sell to your students and clients.

How to Get Started with PLR Content and Make Money

If you use PLR content correctly, you're able to save time and money, while still growing customers, clients, and engagement. You may be thinking how it's possible for you to make money through the use of PLR content, especially because you're paying for it. However, the power of content can make you more financially successful than you've ever realized.

Today's market is all about content, content, content! This makes it crucial for you to ensure you're releasing high-quality content at a frequent rate, regardless of your industry, career path, or objectives. If you're longing to make money online through PLR content, remember that the way you utilize the content is your selling point.

We've established that utilizing good content has many ways of making money, but what are these ways? These following ways of making money through PLR content will answer this golden question (Truex, 2018).

How It Works

As established earlier, digital assets are beneficial for a number of financial reasons. They are extremely profitable due to the small investment you have to make to purchase it. Once you own this digital product, every sale you make is purely profit. This is what can make PLR content beneficial in the long run, regardless of how much you have to pay for rights.

For example, if you buy an online course that teaches people about healthy eating, all you pay for is the rights to this course. You can then sell this product as many times as you want, as there are no more expenses after your first and last purchase.

PLR content is a great opportunity for all entrepreneurs who are looking to expand their business, and grow their income. Regardless of your business, this content can be exactly what you need to further your career. What makes it so useful is how quick and easy it can be! There's minimal work required from you, and the process toward profiting from PLR content is pain-free.

This is what you can expect when you're getting started with PLR content (WayToChanges, 2020):

- **Research.** The most important step to start off with is research. Even if you have a good knowledge of PLR content, it's crucial that you do substantial research! You want to ensure that you're getting good quality content, and you're legally buying the rights to this content. Once you've done your research, you'll be ready to get started with PLR content.
- **Use PLR.me.** When you're looking for a place to purchase your PLR content, it's important to

find a platform reliable and trustworthy. PLR.me is a well-known platform that has a team of writers who produce a variety of content for you to purchase. You purchase your product individually, and from there it's all yours!

- **Choosing your product type.** Although this form of income is easy, it's important to have a clear vision. You want to distinguish the type of products you plan on selling, and which niche intrigues you the most. On PLR.me you'll find products such as eBooks, articles, and checklists, as well as different niches to choose from such as health and fitness and business.

PLR Content as a Marketing Tool

Did you know that your PLR content can be used as a way to market your business? Marketing is an important part of any successful business, so it's vital for you to invest time and money into it. Using PLR content can be a more affordable way for you to market your brand, product, blog, or website.

These are some ways PLR content can help you to market your business:

- **Making an advert.** An advert goes a long way when you're aiming to promote your business.

You can buy PLR content such as videos, audios, or graphics. The video could be stock footage that fits in with the theme of your advert, and the audio could add atmosphere to this advertisement. Purchasing graphics can give you something to post on your socials or on your website that helps to promote your business.

- **Social media campaigns.** We all know that social media is the way of the future. In order to succeed as an entrepreneur, you need to have a good presence on social media, as this is where most of your clients will be based. Being present on social media as a business takes a lot of work and energy. Buying rights to social media content can help you to build a social media campaign that draws a lot of engagement to your socials, thus helping to grow your business.
- **SEO marketing.** Search engine optimization is a great way of writing articles that help you to reach more people. However, SEO can be quite difficult, especially if you lack experience in writing. Purchasing rights to SEO articles can help you to publish blog articles that promote your brand and help you to achieve better engagement.

Using PLR Content for Your Blog

If you have decided to start a blog as a way to make money on the side, then it's a good idea to use PLR content within your blog. You can start a blog without being a professional or experienced writer. There are many articles that you can buy rights to and post on your blog. However, there are a few things you should consider when you are using PLR content as your blog posts:

- **Continuity.** Although all your blog posts should be different, it doesn't mean you should neglect continuity. It's important to post blog articles that are similar and related to each other. This allows your blog page to keep a similar theme throughout its various articles. It may be difficult to find blog articles that are all similar, but you can take articles written on unrelated topics and adapt it to something that will suit your blog.
- **Similar tone and style.** In order to gain a strong audience, it's important to create a writing style and tone that your readers identify with. You'd be surprised as to how much of an affect tone and style has on your reader. If you write sentences and paragraphs in a certain way, you need to keep that consistent. This is relatively easy to accomplish. All you need to do is modify and edit the blog articles you buy

rights to, and you can ensure it has a tone and style that sounds suitable.

- **Keep it trendy.** We all know that blogs have become increasingly popular, and the market is saturated. This makes it crucial for you to keep up with the trends and what's current, so that you can post articles that get more engagement. At the end of the day, the objective is to have a blog that lots of people visit and engage with, so that you can find ways to make a living from this blog. Even though you trust the blog article you bought rights to, you must still do your research and discover whether it's relevant and current.

Another way to use PLR content when you're blogging is by selling some form of content through your blog posts. For example, if you have a blog that's about being positive and finding happiness, then you can sell an eBook that can further help your readers to find happiness and fulfillment. You could also offer courses on your blog posts that are related to the article. Your blog can help to promote your PLR content, and you can use it as a way to sell these products.

Converting PLR Content

Even though you bought the rights to the content, this doesn't mean you have to use it exactly as is. You can adapt it in ways to make it different, as well as more profitable for you! Why not take the opportunity to convert your PLR product into a different version that more people can buy. This can help you to make more money, as you can sell different products that all stem from the same PLR product you bought.

These are some conversions you can do that'll get you extra money:

- **Create an audio.** You can create an audio or video for the content that you've purchased. For example, the PLR content may be an article about how to distance yourself from toxic people. This content can be turned into a quick video demonstrating this with cartoons. You could also convert an eBook into an audiobook, so that it reaches a wider audience. In order to make these conversions successful, you could hire freelancers on Upwork with the necessary skills.
- **Translate your content.** If you're fluent in another language, then you can translate your content in this language. For example, you can sell an eBook in both English and your chosen language. This helps you to appeal to a wider audience and in return you'll get more business.
- **Turn articles into an eBook.** Once you purchase a large number of PLR articles, and you have a huge stockpile of articles, you can

turn that pile into an eBook. This can be achieved by putting all of the articles together into an eBook and selling it as a helpful all-in-one book.

Finding Multiple Streams of Income for Your PLR Content

There are many different ways you can make money through the use of PLR content, but why limit yourself to one avenue? It's worthwhile to use multiple avenues of content to make more money. You don't have to find one avenue that works for you and stick with it, because that will only limit your success and potential.

For example, you could start a blog about something you're passionate about. In this instance, let's say that you're passionate about fashion. You can publish blog articles about the latest trends and fashions and purchase fashion-related PLR content. Once your blog gains popularity and engagement, you can start your own business!

Maybe your dream has been to start your own fashion line, and your blog can catalyse this project. You can buy rights to PLR content such as graphics for your logo, social media content for your social media accounts, and designs for your own clothing. You can then use PLR content to market your business.

This is just one example of how you can use PLR content in various ways to bring in multiple streams of income. Once you get comfortable with the use of PLR content, you can make it work for you!

Making Money by Selling PLR eBooks

PLR content is well-known and popular for being a quick and efficient way to produce content. eBooks generally take a lot of time and effort to create, especially if you're not an experienced writer. However, the financial benefits you can reap from selling eBooks, makes it worth the time and effort.

With the use of PLR content, you're able to purchase the rights to an eBook and sell it as is. This means you have a book with your name on it, without having to spend long hours trying to create it. You can then sell your eBook and watch your investment grow (Noorani, 2019).

Here are some of the best websites where you can find PLR eBooks to purchase and sell:

- **IPDLR.** This website has around 40,000 members, so you're bound to make sales!

- **PLR database.** This is a growing database that has a variety of different PLR content up for sale.
- **Buy quality PLR.** This is one of the most popular sites for selling eBooks that has a large variety of niches to choose from.

When you're looking to sell your eBook to readers who are wanting to read your PLR eBooks, then you should look into using Amazon Kindle. This is a platform that connects people selling eBooks with people who are looking for books to buy and read online.

Chapter 5:

Self-Publishing

Back in the day, publishing a book took a lot of time, money, and connections. You had to jump through a lot of hoops to get your book published. Nowadays you're able to self-publish your own books with ease! If you've ever thought of writing a thrilling story that shocks your readers, or you plan on writing self-help books that help people deal with their stress and depression, your time is now!

Getting Started with Self-Publishing

The journey of a modern-day author is a lot different to how it was in the past. You can go through the whole process of writing a book, editing it, and then producing it, all on your own. This is nice for a number of reasons, one being that it's more affordable. You also get to do things in the way you want it to be. This helps to make a final product that you're happy with.

Getting started with self-publishing is not as difficult as you may think. Anyone is capable of doing it! All you need to do is create a book people would love to read and follow the steps necessary to publish it (Hartgers, 2019).

The Revolution of Publishing

Publishing books is not a new occurrence, as the first printing press in the West was invented by Gutenberg in 1440. As humans, we've always been storytellers and communicators. It was common for people to communicate through written word, as it was a way to reach more people without physically interacting with them. This makes it no surprise that the oldest recorded published book is from the 18th century!

Traditional publishing can be a very long and frustrating process. For as long as publishing has existed, publishers held the power to authors' futures. Publishers would buy the rights to the authors' works and decide whether they were worthy enough to be published. This could be an unsuccessful process for many talented authors who hoped and dreamed that their book would make it.

With the digital age has come many adaptations that have revolutionized the publishing industry (for the better). As an author, you no longer have to rely on publishers to get your book out there, because you are now in control of the publishing process. The online world has made it possible for people to self-publish their books online.

Now writing and publishing books is not only for well-known authors. You can be an inexperienced writer with a message or story to tell. Maybe you have a passion for self-help or business and now's your chance! You have the power to write your own book and self-publish it on your own terms.

You don't need an agent or publisher to believe in your idea; all you need is self-confidence and ambition to get it done yourself. The self-publishing market has become worth $1 billion, which shows how valuable and lucrative this side hustle truly is!

Pros of Self-Publishing

There's a long list of positives that come with self-publishing, as it's a liberating way to get your project out into the world. It proves that the world is becoming a place for independence and entrepreneurial skills, because writing your own book and self-publishing it makes you your own boss. Here are some pros that you can expect from self-publishing:

- **No more gatekeepers.** Unlike traditional publishing, there's no publisher that acts as a gatekeeper to your success. You don't have to go through a middleman who determines whether your book will sell or not, because you get to make that call yourself. The only barrier that can prevent you from sharing your project with the world is you!

- **No time constraints.** When you're an author trying to publish a book, you're required to hire an agent. This agent ensures that you're keeping on track with your writing, and they give you a long list of deadlines you have to meet. These deadlines are necessary because they can affect whether your book gets published or not. When you're self-publishing your book, you can get it done in your own time. It's all up to you and how quickly you need your book published.
- **Creative freedom!** We all know that creativity is a crucial ingredient for a successful book. When you're getting your book traditionally published, you have a lot less freedom to write what you want, because there are gatekeepers who only want books with content they know will sell. This can make it difficult to write a book that is special and fun for you as you're focused on what others want. When self-publishing a book, you can go with the flow and write a book that you'd love to read. By doing this, you're more likely to write a book that will do well in sales.
- **Earning higher royalty rates.** Royalty rates are the percentage of money you get from your sold books. When you self-publish a book, you get a higher percentage of the revenue, and the platform you use takes off a smaller percentage

for providing you with their service. You'll get more money for each book you sell through self-publishing, than you would for traditional publishing.

Cons of Self-Publishing

With all potentially positive and lucrative ways of earning an income, there are cons that you need to be aware of. Unfortunately, there are a few downsides to self-publishing that some authors struggle with, but once you tackle them, you'll be on your way toward success! These are some cons you should look out for when you begin your self-publishing journey:

- **Lack of support.** When you're self-publishing a book, you're doing it all on your own. Even though this can be beneficial, because you have more control and freedom, it can be difficult to know whether you're on the right track or not. You don't have a publisher to tell you if your writing is of a good standard, or whether your book will do well or not. You have to completely trust your thought process, your writing, and your ability to publish a book that will be successful.
- **More start-up costs.** Due to you doing the process on your own, you have to spend your money to ensure that your book is up to

scratch. This means that you should hire freelancers to design your book cover and editors to edit your writing. You could even hire someone to write your book for you! This outsourcing can be costly upfront, but it will help your book seem more professional.

- **Keeping yourself motivated.** As a writer self-publishing a book, you have no agent or publisher rushing you to meet deadlines. Although this can be relieving, because you don't have to rush to get your project out there, it can delay your process. Often that push we get from external people, helps us to stay on the right track. When you're self-publishing, you have to rely on yourself to keep yourself motivated, regardless of the days that make you want to give up.
- **Negative reputation.** Unfortunately, self-publishing has a stigma around it, as it's deemed less worthy than traditionally published books. This is due to the difference in quality in self-published books and traditionally published books. Self-published books are often of a lower quality, as they are independently written, edited, and produced. This makes it crucial for you to spend extra money on professional services, so that you can publish a book of high quality and end the stigma!

Although there are cons to self-publishing, it shouldn't deter you from this avenue of income. All of these negatives are things you can work through and get better at. The more you work on self-publishing your books, the more motivated you'll be to keep on writing. With time you'll care less about the negative stigma and lack of support you have in this industry. You'll also be able to use your income from selling books to put money into your next books you plan on making.

How Much Does It Cost to Self-Publish a Book?

The price for self-publishing a book varies, as it depends on the amount of outsourcing you choose to do. The benefit of self-publishing is that you can spend as much as you can afford to publish it. If you're struggling with money, then you can accomplish the book on your own with little to no professional services. However, if you have money to spend to improve your book, you can choose as many services as you want.

It's a good idea to get a professional graphic designer to design the cover of your book, so it's enticing for potential readers. If you write the book yourself, it's a good idea to hire a professional editor who will modify your book into a mistake-free, well-structured book. If you really struggle with writing or you lack the time to do it yourself, then you can hire a freelance writer to write the book you want. It's as simple as that!

When you're publishing your book on a platform, it may be an added expense, depending on the platform that you're using. Some platforms require start-up funds, whereas other platforms take a commission from every sale you make. At the end of the day, you can spend $1,500–$2,000 on your book, all depending on the services you use during your publishing process.

It may be helpful for you to establish a budget, so that you can clarify what you're going to spend money on. This is how you can create a publishing budget:

- **Editing.** Regardless of whether you are writing your book or a professional is doing it, it's important to invest in some good editing. This will make your book seem polished, professional, and reliable. Paying for editing for a book can cost $300–$1,500, depending on the quality of the editing and the quantity of your word count.
- **Formatting.** There is a certain way your book needs to be formatted so that it can be professional and meet strict guidelines. Most platforms that sell books online, such as Amazon Kindle, require your book to be formatted in a specific way. You can hire someone specifically to do this, or you can find an editor who can do it while they edit. This will cost an extra $50–$300.

- **Book design.** When you're sorting out your book's designs, ensure that you consider both the interior and exterior design. You want a cover that's professionally made, and you may even want different designs in the interior of your book. These designs will cost anything from $100 to $1,000, depending on how many designs you want to include in your book.
- **Marketing.** You can't forget to set aside money to market your book. In order to make the number of sales you desire, it's imperative that you pay attention to your marketing. There are many different ways you can market your book, and you can even find ways to do it for free, if funds are an issue for you. You could pay $0–$500 to market your book.

Where to Get Self-Published

Due to the popularity of self-publishing, there are many self-publishing companies that will help you to self-publish your own book. However, you need to find the best option for you that gets you the best royalties' rates, has the customers you're seeking, and other elements that will help you to become a successful self-publisher. The first thing you should establish is what type of self-publishing company you should use. Here are the three different types of self-publishing companies:

- **Self-publishing platforms.** Using a self-publishing platform can be the easiest way to self-publish your book. These are self-publishing retailers, and it's a great way for you to get into self-publishing, especially if you don't know much about it. Many people getting started with self-publishing choose to use these platforms as a way to get their book published.
- **Educational services.** If you're looking to learn how the process works, then you may want to consider using educational services. These services teach you about the industry, and you're able to learn the ins and outs of self-publishing, without having to rely on self-publishing platforms.
- **Author services.** There are companies that offer professional author's services to guide you toward self-publishing. They'll also teach you how to edit and format your books the appropriate way. This is a good investment, as you can end up spending less money on utilizing freelancers' editing and formatting services. These companies will provide you with self-publishing tools that get you onto the right track.

When you're seeking self-publishing retailers, it's important to choose a platform that's reliable. There are many different platforms out there, but there are a few popular ones that any self-publisher is familiar with. These include the following self-publishing retailers:

- **Amazon Kindle.** Amazon KDP is the most popular self-publishing platform, with over 80% of the market on its platform. You can't go wrong with using Amazon, as there's a huge customer base that are seeking a variety of different niches.
- **Apple's iBooks.** If you have a MacBook and use Apple devices then you may want to consider using Apple's iBooks platform. This platform is growing increasingly popular and has approximately 10% of the market on its platform.
- **Kobo.** This is a smaller platform that is powerful in its own right. This platform has an international market, which helps you to sell from anywhere to anyone. This is valuable, as it opens you up to a wider audience.

You can choose whichever platform best suits your eBook and self-publishing goals. You can try out different platforms and commit to the one that gets you the best results. Many self-publishers choose to publish their books across multiple different platforms to achieve the best results.

This industry is continuously growing and developing. It's said that its next phase will allow authors to deal directly with customers, instead of using a platform as a middleman. eBook subscriptions will become more popular, and the traditional way of publishing will be less sought after. The stigma of self-publishing will slowly fade away, as people realize self-publishing is the way of the future.

How Self-Published Authors Have Succeeded

Everything you've heard so far may sound very convincing, but you may still be wondering if it's possible for you. Many individuals have experienced amazing success stories from self-publishing, and you are more than capable of being the next story! There are countless authors who are earning over $50,000 from their royalties, as well as authors making a six-figure income through self-publishing. Here are some success stories that can inspire you to get started as a self-published author (Haysom, 2020).

Holy Island *by L. J. Ross*

L. J. Ross was a lawyer who decided to pursue her passion of writing by self-publishing a series of books.

She's published a total of 19 novels on Amazon Kindle, and she has experienced success from them all. However, one book that stood out the most was *Holy Island*. This novel was a perfect mix of romance, crime, and suspense, which many critics questioned, but its success rate proves that it works.

Only the Innocent' *by Rachel Abbott*

Many self-publishers take time to reach large success, but luckily for Rachel Abbott, she quickly gained success from her first book, *Only the Innocent*. This book is about a woman who is a murderer, which helped to gain all of its hype. It's not common to hear stories about women being cold-blooded murderers, as they are often men. Abbott's unique take on a very well-known concept helped her gain her fast success in this industry.

Horror books by Adam Nevill

Horror is the genre we all love and hate at the same time. Adam Neville is a professional writer who specializes in horror. Unlike the other success stories, he originally started off with traditional publishing, with 19 books published under his name. He's won awards and had his novel *The Ritual* adapted into a movie on Netflix. It's fair to say that Nevill was already a successful writer before he started his journey with self-publishing.

He started with self-publishing so that he could have more control over his projects, as well as reap the financial benefits. The first novel he self-published was *The Reddening*, and the sales surpassed the sales of most of his traditionally published books.

Tips for Success

Anyone is capable of succeeding when self-publishing their books. You could be the next successful self-published writer who changes the game! However, there are dos and don'ts, as well as tips that you must follow when you're trying to create a successful and sustainable career in self-publishing. These are a few tips you should consider when you're entering this industry:

- **Invest the necessary time, money, and creativity.** Did you know that around 30% of Amazon's bestselling books are self-published? Self-publishing is growing increasingly popular, meaning there's a big market for them and potential for you to make a lot of money. The downfall is that there's a lot of competition, as there are many other people on platforms doing the same thing. This makes it crucial for you to invest your all into your books, so that you can stand out from your competition.
- **Budget efficiently.** Although self-publishing is more affordable than traditional publishing,

there's still a lot of costs that go into self-publishing the perfect book. In order to make a book that is professional and enjoyable, you need to pay extra money to make it up to scratch. It's important to create a budget before self-publishing a book, so that you can have realistic expectations for your book. Once you make revenue from your self-published book, it's important to put a good percentage of your profits aside. This money can be used to make your next book even better and more professional.

- **Host a launch party.** Promotion is a crucial part of self-publishing a successful book. There are many different ways you can market and promote your book, and hosting a launch party is a popular way of promoting your book. A launch party is the act of gathering your launch team, and getting them to write reviews on your book. Reviews are valuable when you've released a self-published book. These reviews will help Amazon's algorithm to get more people to view your book, which will result in more purchases. Having positive book reviews also convinces people to buy your book, as they get a different perspective on the contents of your book.

- **Be passionate about your book.** Above all, it's important to have a passion for the book that you're writing. Passion helps any project to become a success. You should write a book about something that excites and interests you, as this will help you gather the time and energy to work on it every day. Pick a niche that's down your alley, for example you may be into health and fitness, so you can choose to write a book about dieting and working out. Ensure that your book represents something you're proud of, because this will make you excited to promote and sell it!

Chapter 6:

Affiliate Marketing

Maybe selling products isn't for you. You don't want to take the time to make a product that can be sold through a platform. If this is you, then you may want to use affiliate marketing as an avenue of making income online. As an affiliate marketer, you promote other businesses' products. This is a simple way of earning a passive income without having to sacrifice a lot of time and energy.

What Is Affiliate Marketing?

If you've done your research about passive income and making money online, then you've probably come across the term "affiliate marketing." If we break down this term, we can determine that affiliate means to collaborate or be connected with an organization. Marketing is a way of promoting a brand, product, or service. Once understanding this, it's easier to understand affiliate marketing as the act of collaborating with a company by promoting their product and earning commission in return.

Affiliate marketing is a popular side hustle for many digital creators, as it doesn't require a lot of time or money. It can also provide you with a respectable amount of money each month, from a couple of hundreds to even thousands of dollars some months. It all depends on how seriously you take affiliate marketing. These are a few things to remember about affiliate marketing before you get started:

- **Commission.** You get paid a commission as an affiliate marketer, meaning that you only get paid once someone purchases a product through your link. Affiliate marketing only works when there are customers, as their purchases result in the revenue and commission that is received by the merchant and the marketer. This is why it's important for the marketer to use a platform that has the right target audience.
- **Building traffic.** In order to get a consistent customer base from affiliate marketing, you need to build traffic, and this can take time and energy. On platforms like Amazon, building traffic means that you get a frequent number of viewers on your promotion, resulting in more sales. You need to consider a variety of things when you're aiming to build traffic, but the most important thing to remember is to remain consistent.

- **Be open with your audience.** As an affiliate marketer, you're required by law to disclose information to customers purchasing through you. Your audience needs to be aware that you're making money from their purchase. Although this may seem like a detriment to disclose, it can help your business. By being open and honest with your audience, they're more likely to trust the product that you're promoting, which influences them to purchase it.

Pros of Affiliate Marketing

Becoming an affiliate marketer has its perks. If you're looking for a way to break away from that 9–5 job that's draining you, getting into affiliate marketing will get you two steps closer to that dream. Affiliate marketing is a good place to start when you're aiming to become a successful digital creator. These are some of the pros that come with being an affiliate marketer:

- **Requires less work than running a business.** You may be someone who's always dreamed of selling their own products, but you've never had the energy or confidence to start your own business. By becoming an affiliate marketer, you get the opportunity to sell products without having the pressure of starting your own business. This can even serve as a test run that helps you to establish whether starting a business is something you'd like to do.

- **Anyone can do it.** You don't require any expertise or skills to get started as an affiliate marketer. You just need to be open and eager to learn a new way of making money online. Fortunately, this avenue of work takes practice to become perfect, and many people start their journeys off by learning as they go. There's no pressure or expectations for you to be perfect when you're getting started.
- **Low cost.** What makes affiliate marketing a good starting point for digital businesses is how low cost it is. You have few to none upfront fees to pay when you get started, so if you don't end up succeeding, it's not a huge loss. This makes it a lot more exciting, because stakes are lower, and all the money you make from commission is purely profit. You'll have a few expenses when you're first getting into this business, but if you do enough research, you'll find the most affordable and reliable ways to do things.
- **Flexibility and independence.** We all want to leave our 9–5 jobs, because we're tired of the lack of flexibility and independence we get from these jobs. When you're an affiliate marketer,

you essentially become your own boss. You decide the marketing tactics you use, and how you'll get customers to purchase through your link. This can be freeing to have control over what decisions you make. Affiliate marketing also allows for flexibility, as you get to choose your own working hours.

- **It's a form of passive income.** Passive income is a form of income that you can make without actively working on it. When you're getting started with affiliate marketing, you need to be very active and consistent. This will help you to build traffic, sufficient customers, and a following. Once you've taken the time to get into affiliate marketing, and you understand the ins and outs of this trade, you'll be able to make money from it in your sleep! There's no better feeling than waking up and seeing money you made overnight from your side hustle.

Cons of Affiliate Marketing

Unfortunately, there are downfalls that come with affiliate marketing. It's a competitive, unreliable field that can take some time to get used to. This is why it's advisable to do affiliate marketing as a side hustle, rather than a full-time job, because you can't be guaranteed a steady monthly income. Here are some cons you may experience while working as an affiliate marketer:

- **Unreliable source of income.** It's no surprise that earning commissions by selling someone else's products can result in an unreliable source of income. Regardless of how much work you put into your marketing, you're going to have months where few people engage with your link, and that's okay. One month with less revenue, doesn't mean that you're a failure at this industry. You just have to be prepared to wait out the drier periods, because you'll have months where your bank is flooded with commissions!
- **Highly competitive.** With any easy way of making money, there's going to be a lot of competition. Due to the fact that anyone can become an affiliate marketer, there are large numbers of them on different platforms. You may find that there's more competition in specific niches, so you could try avoiding these niches to lessen this competition. Unfortunately, you can't control this competition, as there's very little that you can do to stand out from other affiliate marketers. You just have to hope customers get to you first!

- **Freelancing is difficult.** Although it's a benefit to have a freelancing job that you have control over, being a freelancer is not always glamorous. Unlike with normal jobs, freelancing is very independent, and this can be lonely and daunting at times. You don't have colleagues and a boss who can provide you with words of encouragement and validation. This means that you have to have the willpower to get up every day and resist the urge to give up when things aren't going your way.
- **Lack of repeat customers.** Having a strong customer base is an important part of any successful business. You want repeat customers who love your products to continue buying from you, but this doesn't happen with affiliate marketing. Once a customer has purchased through you, you're unlikely to have them come back as repeat customers. They've bought a product through you, from the merchant, and this means that they'll go straight to the merchant next time they want to repurchase the product.

How to Become an Affiliate Marketer

Anyone is able to become an affiliate marketer, as it's a fairly simple process. You don't need any specific qualifications or experience, as long as you're prepared to work and learn. As an affiliate marketer, you're your own boss who works from home, so how you go about starting this process is entirely up to you. You should observe your own progress to notice whether you're improving in the industry.

Affiliate marketers are all different, as there are a variety of ways you can go about doing it. It's important to choose a way that you're most comfortable with, because this will ultimately result in more sales through you. You should also consider what type of products you'd like to sell through affiliate marketing. It's important to remember at the end of the day that you should sell products you agree with. These are some different ways you may consider to get into affiliate marketing:

The Blogger

If you're an online side hustler, chances are you've created a blog. You may be posting blog articles as a way to generate some more income. If you're a frequent poster with good engagement and a substantial following, then it would be the perfect way for you to get into affiliate marketing. This is how you can integrate affiliate marketing into your blog posts:

- **Find your niche.** When you start a blog, it's important to find a niche. You can post a wide variety of different articles, but you need to ensure that they all fit the same theme. This niche should be something that you're interested in and passionate about, as it'll make it easier for you to maintain it.
- **Sticking to your brand.** Once you're posting articles that reflect your niche, it's important to get into affiliate marketing that is apt for your brand. If you're a blogger who teaches people how to get into makeup, then you should promote makeup-related products. This way you're promoting products to the right market, and you'll experience a higher success rate.
- **Integrating products and courses that reflect your brand.** Regardless of how popular your blog is, you need to integrate your product placement in a way that's enticing for customers. For example, if you're writing an article about how to look after your skin when you do your make up often, you can promote a skincare product. This is relative for what you're talking about, and it could be useful for people who are reading the blog post for skincare tips.

You don't have to have a well-established blog to use it for affiliate marketing. If you're just starting off, or your blog hasn't seen results yet, you can still utilize it. The key to growing your blog is becoming consistent and creative. Once you post articles that people want to see, you'll start to get the engagement that results in multiple purchases through your link.

The Reviewers

Reviewers can be the most reliable sources for consumers. Nobody enjoys buying products without reviews and descriptions, because what if they end up wasting their money on a product that's not worth it? This is why starting a page as a reviewer is a great way to get into affiliate marketing. You get to provide a service of informing consumers, while collaborating with brands who want to get their products out there.

When becoming an online reviewer, your reputation is fundamental. In order for you to become a reliable and trustworthy source that influences consumers, you need to prove to them that your reviews are true and authentic. This means that regardless of your affiliate marketing deal, you need to write truthful reviews.

If you try the product and have a bad experience, you should avoid promoting it to your readers. But, if you try a product that you love, you can express that to your readers, and this may influence them to buy it through your link. If you write a good review on a product that isn't of a good standard, consumers will write you off as a reliable source.

Becoming a reviewer is another good side hustle that can bring in a good income. It may sound simple to try products and review them, but there's more to consider when you want to be a successful reviewer. These are tips that will help you to become a successful reviewer:

Be Wary of Your Language

As a reviewer, the language you use in your writing is what persuades or deters customers from reading your review. If you're writing a positive review that you want people to purchase, you should use emotive and persuasive language. This can make the readers aware of how great your experience with the product truly was.

You should also take note of the type of consumers who will be reading this review, as you should adapt your tone and style to their taste. If you want to make your name known as a reviewer, you can add a phrase or term that you use frequently. This way your personality is portrayed through your reviews, and this can make you appear more reliable and trustworthy.

Do Your Research

Merely writing a review based on your experience with the product is not enough. You need to do research that gives you a holistic idea of what you're reviewing. This means you should research other people's experiences with this product, as well as view the company's competition. Are there other companies selling the same thing at a more affordable price? Have customers had better experiences with the competition? All of these elements can help you to write a fair and true review that accurately represents the product.

Use constructive criticism

If you have a negative experience with a product and you want to write a bad review of it, then it's valuable to provide constructive criticism. Instead of only complaining about everything, add criticism that you think would improve your experience. When you're writing a review for affiliate marketing, you don't want to bring down the product you're promoting, because then no one will purchase it through you.

However, the product you're promoting may have issues you're not happy with, and you want to express your honesty. You can go about this by expressing all the positive aspects of this product, followed by constructive criticism that could improve it. This may even result in more sales, because consumers trust your honest approach.

Be Consistent

When you're trying to establish your name as a reviewer, it's crucial to be consistent. You want your readers to view you as a reliable source for reviews, so that they can follow your advice. If you're a reviewer who's an affiliate marketer, your positive reputation will help you to get more sales through your link, as people trust your consistency.

To be consistent you need to write honest reviews that truly depict the product. This means that you should pay special attention to your research, so that you're filling all of your reviews with informative and relative content.

The Influencers

We all know that social media platforms have become the best asset to utilize as a business. There are countless apps that people use to stay connected and up to date with trends. It's no surprise that social media is a great platform to use as an affiliate marketer, as many companies want the products to be promoted by social media influencers. An influencer on social media is someone who has a substantial number of followers, strong engagement, and an audience who is influenced by them.

You may be wondering how an individual can have an influence on people through a phone screen. People online with a strong social media presence have the power to influence and inspire others, just as much as well-known celebrities do. Anyone can become an influencer, but it takes time and dedication to achieve.

How to become the influencer that businesses want to collaborate with:

- **Be yourself.** Social media often has a bad reputation for being fake, as many people choose to show a false facade online. Although people think that followers want to see you living a lavish, perfect life, this couldn't be further from the truth. Yes it's important to showcase the highlights of your life, but it's also important to show people your reality. You need to show your followers that you're a normal human being who experiences both good and bad days.
- **Find your message.** It may sound cliche, but every individual has a message they want to bring into this world. If you're someone who's experienced bullying growing up, and your motive is to spread love and kindness, then you can make this message known! Having a beautiful message that people stand by makes your followers more comfortable with you. They know you have a kind heart, and this will help them to build a connection with you.

- **Form a relationship with your followers.** The most crucial part of being an influencer is being able to build a relationship with your followers. This relationship will make them support you and your marketing ventures. Building a relationship with your followers is easier than it seems, because all you need to do is be real and raw. Post about your personal life and what you're doing on a daily basis, as this makes your followers feel included in your life. Showcase your personality and sense of humour, so that your followers get an accurate perception of your true nature.
- **Post frequently.** The best way to grow a following on social media, especially on Instagram, is by posting frequently. For many social media platforms, there are different forms of posting you can do. For example, Instagram has stories and feed posts. It's important to post lots of different content on your feed that you know your followers would love to see. Posting your life updates on your story is a great way to showcase your personal life. When you post frequently, the algorithm boosts your views, and this can increase your followers and engagement.

Once you become an influencer with a strong presence online, you can use your account to do affiliate marketing. You can reach out to brands, or brands will reach out to you for collaborations. You may even get a discount code that your followers can use to purchase through you.

When you're deciding to collaborate with a brand, you must consider whether it's something you feel comfortable with representing. Work with businesses that suit your identity and message on social media, as this is your target audience. It's important to represent brands that match you as an influencer, as this can enhance your image.

Affiliate Marketing Through Amazon

Amazon has an affiliate marketing program that can help you to get started as an affiliate marketer. If you're a website or blog owner, and you're seeking for a way to monetize it, then all you need to do is sign up with Amazon Associates for free! They will add affiliate links, that are linked to products on Amazon, to your website or blog, and once people purchase through them you earn a commission.

Amazon Associates will provide you with a step-by-step guide that will get you started. The benefit of using Amazon as the middleman for your affiliate marketing, is you don't have to worry about organizing collaborations with companies. They will take a percentage of the commission as a payment for their services. There are many other platforms that provide this service, but Amazon is one of the most well-known platforms that will have never-ending Amazon listings to use.

Tips for Success

Becoming an affiliate marketer is straightforward and accessible to anyone, and this is why it's such a saturated market. You can get started anyway you like, as these are just a few common ways people choose to do affiliate marketing. Once you get familiar with the process of affiliate marketing, you must work on improving your craft so that you can reach your desired monthly income from commissions. These are tips that affiliate marketers used to receive their best results (https://www.facebook.com/adam enfroy dotcom, 2018):

- **Spend time on your content.** Many people getting into affiliate marketing make the mistake of producing low-quality, fast content. Although affiliate marketing is a low-maintenance job that doesn't require a lot of your time and energy, you still have to spend time on producing quality content. Whether your content is Instagram pictures, blog articles, or reviews, your content has to entice people, so it persuades them to purchase through your link.

- **Monitor your progress.** As with anything, it's crucial to continuously track your progress. You should be mindful of the different things you try out with each post and monitor the outcome of these changes. Once you discover all the things that work and don't work, you can formulate a method of doing your affiliate marketing that's successful for you. Affiliate marketing can be different for each individual, and what may work for your Instagram account, may not work for a blog. The data that you collect will help you to avoid things that don't work for you, and you can utilize the strategies that bring you success.
- **Be useful to consumers.** A great way to become successful at affiliate marketing is by being useful to your consumers. Create content that is informative and useful for them. For example, if you're a blogger selling an herbal drink that helps insomnia, you can write a blog article about how sleep is important for our brains to function efficiently. Along with mentioning the product as a tool to aid sleep, suggest other things that'll prevent insomnia. This will be useful to readers and can persuade them to try the product out for themselves.

- **Lower your expectations.** It's true that you can make a lot of money through affiliate marketing, but you can also have an experience where you make little to no money. Your income is purely reliant on paying customers, and you won't always be guaranteed to get customers. This is why it's important to have reasonable expectations, as it'll prevent you from being disappointed and giving up. You may experience a lucrative month filled with sales, and the next month no one buys through your link. Remember this business is rarely stable, and it's important to keep going regardless of the dry months.

How to Become a Merchant

Maybe becoming an affiliate marketer doesn't fit you. If so, then you can become a merchant. Merchants are the businesses who promote their products through affiliate marketers. You don't need to be a big corporation in order to utilize affiliate marketing, as it can be used by all the different types of businesses!

If you're a company looking to get your product promoted, then you could consider becoming a merchant for affiliate marketing. As the merchant, you're able to find affiliate marketers who will promote your products. The benefit of using affiliate marketing as a way of marketing your products is that you don't have any start-up costs. You don't pay the affiliate marketers upfront, as you only pay commission to them once they sell a product through their link.

This is a step-to-step guide that will help you to get started as a merchant for affiliate marketing (Patel, 2015).

Step 1: Thinking of the Perfect Product Idea

In order to become a merchant, you need to have a product to sell. This is why it's important to spend time thinking about the perfect product idea. Believe it or not, coming up with a product to sell is a lot easier than it seems. It doesn't have to be something completely unique and out of the box. It just needs to be in demand.

It's important not to choose a product that is too close to your heart, as this can make it too specific and difficult to sell as an affiliate marketing product. However, that doesn't mean you can't choose a product you're interested in. In fact, it's a good starting point to think about what you need. What's something that you would buy that you're in need of currently?

Most of the time, when there's a product you need, other people are in need of the same thing. You can look around your home, think of your daily activities, and gather what's missing that you could find in a product. For example, you may be someone who uses their phone a lot throughout the day, but you find that your battery doesn't last you the whole day. You struggle to get it charged, but you still need it for your important phone calls. A product that could solve this issue is a power bank.

Once you've come up with your product idea, you should do some research to see what the competition looks like. In this instance, you'll search for power banks and see what other power banks being sold have to offer. From this you can determine what your power bank will provide that will make it different. You could sell a power bank with a longer battery life, or design it in a cool way that makes it stand out from its competition.

In your research, you should also discover what issues people may have with the product. This can help you to draw up a pros and cons list that'll help you to determine whether this product is a good idea or not. If you notice that the cons outweigh the pros, then you may want to consider choosing an alternative product idea; but if the pros outweigh the cons, it's a good plan to follow through with your idea.

Step 2: Check the Market for Your Idea

After coming up with an idea that you're satisfied with, it's time for you to determine whether there is a market for it. There's no point in selling a product no one is willing to buy, as you'll end up with few to no sales. This makes checking the demand for your product the best way to validate it.

You can find out if people would buy your product in a number of ways. You could ask friends and family whether your idea is good, or you could venture on social media and ask for the public's opinion. However, the only problem with doing this is that you may get inaccurate responses. The people you know are more likely to sugar coat the truth and tell you that they like your product, even though they wouldn't spend any money on it.

This is where platforms like Topsy come handy. Topsy is a tool where you can search URL links and discover how many people tweeted it. If you're searching power banks and find a listing that looks popular, you can copy this link and paste it on Topsy. You'll then see how many people tweeted about it, giving you a rough estimate of the people interested in the product.

You can then communicate with these people and ask whether they'd buy your product. Be sure to ask whether they would actually purchase it instead of asking whether they merely like your idea. If you get a lot of positive responses, chances are you'll be able to successfully sell your product.

You can also do further research to determine whether your product idea is something that's in demand. Often when a product has lots of competition, it's something that many people want to purchase. So, instead of letting competition scare you off, use it as a tool to improve your product so that you can stand out!

Step 3: Creating Your Product

If you're new to affiliate marketing and you're planning on releasing your first product, you should consider selling a digital product first. This will make it easier for you to get started, and it will give you the space to learn the industry as you go. Selling physical products can be more complicated, as you have a lot more to organize.

You can sell almost any digital product through affiliate marketing. These are some popular digital products that get sold through affiliate marketing:

- **eBooks.** As stated earlier, eBooks are a popular digital asset that is in high demand. It's important to write a book that fits a good niche that could be used to create content for affiliate marketing.
- **Podcasts.** Podcasts are becoming an increasingly popular form of entertainment and information. People can subscribe to your

podcast and listen to it while they go about their daily activities.

- **Courses.** Courses will always be in demand, because there's a long list of people who want to learn new skills, hobbies, and career paths.

You can't go wrong with selling digital products through affiliate marketing. All you need to do is discover the demand and how you can meet it. Most of these digital assets take time to make, but once they're made they're quick and easy to sell. You can also use PLR content as a merchant.

Step 4: Find Affiliate Partners

Once you have your product, you need to find affiliate partners to work with. Luckily for you, there is an abundant number of affiliate marketers looking to collaborate with merchants. The best way to go about affiliate marketing as a merchant is to use a program to facilitate the process. Tools such as Gumroad and Digital Product Delivery can help you to get started.

Finding the right affiliate for your product can be difficult. You may have eager affiliate marketers who are looking to collaborate with you, but their niche doesn't suit your product. It's vital to have affiliate partners who have a similar niche and audience to you, because this will get you the sales you desire.

Does Affiliate Marketing Work?

With any way of making income that's easy and convenient, it can sound too good to be true. Many people are deterred by avenues of income that are quick, low-cost, and financially beneficial. There has to be a catch, doesn't there? However, affiliate marketing is as good as it sounds! It's a convenient way of making money without having to start your own business.

You can try out becoming either an affiliate marketer or a merchant and from this decide what role suits you and your lifestyle more. Using affiliate marketing as a business is a great way to make sales and expand your customer base. However, affiliate marketing doesn't always work for all businesses. Your business may not be easy to market in this way, and consumers aren't persuaded to purchase via links.

As an affiliate marketer you will experience days with a lot of commissions and days where you struggle to make any sales through your link. Once you get the hang of affiliate marketing, you'll understand what works for you and get into a routine that rewards you with the best results.

Although this form of income is not always reliable, it's still an effective way to make money. You can experience real success with affiliate marketing if you use the necessary tips for success such as staying consistent, building a brand, researching your partners, and following legal requirements. If you want to be even more successful in this industry you can use variety.

Using a variety of different advertisements for affiliate marketing will help you to keep your readers interested. If you post the same ads with similar content each time, it can become monotonous, causing your audience to ignore these ads. You can show variety in the following ways (Duermyer, 2020):

- **Switch up your content**. You may be a blogger who does affiliate marketing. This is a good opportunity for you to switch up your content. You can post blog articles that are unique and interesting, as well as structure your pages in an eye-catching way. This will keep your audience engaged and interested in the content you post. This will influence them to check out your affiliate marketing links, and this can lead to more purchases.
- **Promote different merchants.** Although it's important to partner with merchants who are suitable to your brand, you should promote a variety of different products. Be sure to partner with businesses that are appropriate to your brand, but also provide something fresh and interesting to your audience. For example, if you're an influencer who posts content about makeup, you can partner with a skincare brand.

This is a good product to sell, and it shows your audience that you not only care about makeup, but you also advocate for good skincare.

- **Utilize an email list.** An email list can be a very useful tool as an affiliate marketer. You create an email list by getting people to provide you with their email address on your chosen platform. This gives you the access to email them, and inform them of your latest updates. In these emails you can include affiliate marketing links that are relevant to the content in your emails.

Utilizing these effective marketing strategies mentioned in this chapter will help you to see results in your affiliate marketing. If you're still struggling and you want some extra help to find the success you desire, you can join an affiliate program that aids you along your journey in affiliate marketing. A good affiliate program to look into is BigCommerce, as it can provide you with the following:

- **Ready-made links.** Once you have already made links and texts you need to use for promotions, it makes your life a whole lot easier. All you have to do is create content that will be used to promote the link.
- **Support from your account manager.** When you sign up for a program like BigCommerce, you get an account manager who is experienced

in the affiliate marketing business. They'll have advice and tips that can help you through any troubles you may face.

- **Insightful tracking.** Their platform helps you to track the success rate of your links. You can see useful insights such as the performance of your links and how many sales you made, and you can get your money at the same time every month.
- **Powerful marketing strategies.** With all of BigCommerce's knowledge and insight, they are able to use strategic moves that will grow your page. You'll find that you get more engagement and a higher rate of people purchasing through your link.

Common Affiliate Marketing Myths

What deters many businesses from getting into affiliate marketing is their false perceptions of it. These perceptions are often created by myths that are commonly associated with affiliate marketing. These myths also prevent people from pursuing work as affiliate marketing, and this can be a misfortune for many to miss out on this opportunity. These are some myths in this industry that shouldn't deter you from pursuing affiliate marketing.

A Quick, No Maintenance Process?

Many people think that you can make fast money from affiliate marketing by simply getting started. Although it's true that you can make good money from this industry, it's not as quick and easy as people think. This myth can lead to high expectations that create disappointment when you don't reach those heights.

It's important to know that affiliate marketing does take work and maintenance, especially when you're getting started. It takes time to make quality content and build traffic and engagement. This can make it a long process until you start making money through your links. It's fundamental to stay consistent and patient in order to see the results you long for.

Once your affiliate marketing picks up, it doesn't mean you must leave it and expect to keep making money consistently. You need to constantly monitor and work on it in order to keep up your positive results. Maintenance of engagement is less work than starting from scratch, but it's important to continue making engaging quality content, while utilizing your marketing strategies.

Only Use a Popular Niche?

Although it's important to use a niche that is sought after by customers, it doesn't mean you're only limited to what's popular. You should venture out to different niches and choose something that piques your interest. There's no point in selling products of a niche that doesn't excite you, because it may negatively impact the quality of what you produce. As an affiliate marketer, you should focus on niches that represent your brand and not only collaborate with partners of popular niches.

If you're a unique business looking to get into affiliate marketing, don't be deterred by what's popular. Stick to your brand and partner with affiliate marketers who will represent you best. You never know—your product could be the next big thing in affiliate marketing!

Products on Multiple Sites Equals Success?

Many merchants aim to have their product links on as many sites as possible, and this is not the way to go. With affiliate marketing quality over quantity will always get you the best results. If you decide to get your product links promoted by many different affiliate marketers, you may not get the results you want. This may be caused by partnering with the wrong type of marketers who won't have your target audience or marketing strategies that'll boost your link.

When you spend time on finding quality affiliate marketers who will best portray you and your brand, you're going to get better results. You'll want marketers who portray your brand, as they'll be advertising to the right target audience. You should also seek marketers who use social media and SEO as a means to promote your link.

Chapter 7:

Coaching and Consultants

Have you ever thought of yourself as a teacher with knowledge and skills? Do you wish to change other people's lives by sharing your talents and expertise? Are you interested in the idea of sharing valuable information to people who are eager to learn? If you answered yes to any of these questions, getting into coaching or consulting may be a suitable side hustle for you!

Why Coach and Consult?

Do you have what it takes to make a good coach or consultant? You may be thinking that you're not the perfect candidate for this role, but you'd be surprised as to how good of a coach or a consultant you could be! There is a lot to coaching and consulting that can make it both a rewarding and exciting job. You first need to determine what coaching and consulting is and whether it's suited for you before you continue.

What Makes a Good Coach?

You may have never considered getting into coaching before, because it's nothing you've had a great interest in. You may even think that you're not suitable to become a coach, as you have no experience in teaching others. However, some of the best coaches, who have coached countless individuals to their successes, stumbled upon this job by accident. This is why you should take the time to look through what qualities and skills make a coach, so that you can determine whether it's for you or not.

Here are some qualities and skills that make a good coach:

- **Open-minded.** A valuable quality to have as a coach is open-mindedness. When you're planning to share your knowledge and skills with others, you need to be able to be open to different ideas. Although you're getting into coaching with the objective to teach what you know, it's vital to be open to what others know and how they've learned how to do things. Incorporating other people's ideas and experiences can often improve your coaching and their experience with you.
- **Goal driven.** As the coach, you're essentially the person in charge. Not only do you have to worry about being on track with your coaching,

but you also have to monitor whether your clients are on track as well. In order to achieve this, you need to be a goal-driven person who sets goals and works toward reaching them. Being goal driven will help you to become a consistent and reliable coach, and you'll be able to motivate your clients to reach their goals.

- **Observational skills.** If your aim is to be a good coach, it's important to be observant of your clients. You don't want to teach by speaking at them, but rather you want to provide them with an experience where you're hands on. Being observant helps you to become an active coach who can notice clients' progress and challenges. These forms of lessons will help clients learn better and see the results they hoped for.
- **Deliverance of constructive criticism.** It can be daunting to tell a client when they're doing something wrong. You don't want to embarrass them or crush their spirits, but it's important for them to know. Without constructive criticism, your clients will never know what they're doing wrong, and this can make it impossible for them to improve. This is why it's important for any good coach to learn how to deliver constructive criticism that teaches clients in a polite and friendly way.

What Makes a Good Consultant?

If you're knowledgeable and have expertise in a specialized career path, then consulting is perfect for you. Many consultants are people who branched out in their profession by helping others in the same field reach their career goals. For example, someone who is a consultant may choose to spend their spare time consulting for accounting firms that need help or for businesses that need financial advice.

Consulting can be more than just giving advice to clients, as it can offer so much more! Consultants provide clients with valuable information that they can utilize in the future, and this can improve their careers. They can also solve the clients' problems by giving them solutions that they can implement. These are some skills and characteristics that would help you to be a good consultant:

- **Problem-solving skills.** As a consultant, one of the most valuable skills to have is problem-solving skills. You'll often be confronted by clients with a variety of different issues. This can make consulting a difficult job, because you can never predict what problems you'll be faced with. Having good problem-solving skills will help you to determine what the issue is with all the problems you're faced with from clients.
- **Resourcefulness skills.** Depending on the job and the clients you have, you may have a wide

variety of problems that require lots of different resources. For example, you may be an IT consultant and your client needs a specific hard drive that you don't have. This is when you have to become resourceful by making plans to achieve the results your clients want. Being resourceful means that you can make a plan with the existing resources that you have.

- **Good communication skills**. As a consultant, you have to communicate with your clients about what their problems are, and from this you have to investigate. Listening is an important communication skill to have, because if you don't actively listen to what your client wants, you may be lost about what your assignment is. Once you're done investigating and doing what you need to do, it's time for you to communicate to the client again. You have to let them know the problem and the different solutions they should implement.

Benefits of Online Coaching and Consulting

Online coaching and consulting have their differences compared to traditional coaching. Many of these differences make it an easier and more convenient way of coaching and consulting. There's a lot of positivity and benefits that could come from pursuing this as a side hustle. You'll end your months with more cash in your pocket and a more enriched mind. These are some benefits of getting into online coaching and consulting:

- **Learning more about your craft.** In life, we can never stop learning, because there's always something new to discover, especially in our passions and interests. When you're teaching others about something you're passionate about, you'll find yourself learning new things you never knew. As a coach or consultant who has to physically practice your craft, you'll notice that you get better at what you do.
- **Feeling rewarded.** There's nothing more rewarding than feeling like you've helped someone. When you're a coach, you have the opportunity to help people learn a new skill, or you can help them to improve at something that's valuable to them. This could potentially change someone's life and make them well-equipped for success. As a consultant, you can help clients to get past issues they've been struggling with. This can help them pursue a successful career and become better at dealing with these problems when they're confronted with them in the future.
- **Making some extra cash.** Why not make cash off of knowledge and skills that you already

have? Not only can you help people learn, but you can also make a decent profit from it. Getting started with coaching and consulting is low-cost, making the majority of the money you make your profit. There are many people out there looking for a more affordable way to learn necessary skills and solutions, and you can provide this service at a cost that is reasonable.

- **Growth in confidence.** We all have doubts in our abilities at times, no matter how skilled and talented we are. When you teach your craft to others, it can always help you to build confidence in yourself and your ability. Knowing that you're skilled enough to share your craft makes it easier to believe that you're truly good at it. You may even find yourself improving these strengths or reaching new heights in your career, because your confidence gives you the boost of energy you need to excel.

Coaching Through Social Media

Social media has made it even easier for coaches to reach an audience. As a coach, you no longer need to physically seek clients or spend lots of money on elaborate advertisements because all you need to do is utilize a social media account. For many coaches, Instagram has become a great platform to start their coaching hustle. Anyone is capable of becoming a coach through Instagram, as long as they follow the social media tips that will get them the number of clients they aim to get (Morgan, 2014).

What Your Followers Want

When you take a look at your Instagram account, you need to consider a variety of things. What is your brand or image that you portray online, and what do your followers follow you for? For example, you may be someone who's into fitness, and you post a lot of your fitness journey online. Your followers may follow you, because they love your lifestyle, and they want to know how they can get as fit and healthy as you.

If this is what you represent, then you may want to consider becoming a fitness coach, as it's something you're passionate about, and you know your followers associate you with fitness. This will be your niche, and you can work toward making a fitness program for your potential clients.

When you're working on your fitness program, it's important to consider what your followers want. While doing this, you should determine the following:

- What gender are you targeting?
- What age group are your potential clients a part of?

- Are your potential clients looking to lose lots of weight, or do they want to tone their bodies?
- What types of exercises and eating plans would your potential clients be interested in trying?

A great way to get answers for these questions is by directly asking them and posting polls on your story. Asking these questions will give you a better idea of what your fitness program should look like. Once you determine your niche, and what your followers would want you can create a coaching program that would be effective and helpful for anyone purchasing it.

Share Your Story

The best way to make yourself credible to your audience is by sharing your own story. When people find out your personal journey, it helps you to be seen as a more reliable coach. You need to think about any incidents in your life that brought you the talent or skills you're teaching to others.

For instance, if you're becoming a fitness coach, think about what happened in your life that motivated you to start your journey. You may have been dangerously overweight and experienced health risks, so you decided to change your life around. You started eating healthy and followed a strict workout plan. This story can inspire many to follow in your footsteps, as they'll want to be coached by someone who made it happen for themselves.

These stories make you a more credible source, because people can see that you're actually good at what you do, and this can help them to achieve what you've achieved. Sharing your stories in life can also personalize your Instagram account by showing people the reality of your life and how you got to where you are now. This helps to build a relationship with your audience, as you see yourself as a person, rather than a stranger on the Internet.

You should also engage with your followers when they comment and ask questions. Let them know that they're seen, because this lets them know that you'll be an attentive and hands-on coach. This can convince them to use your services.

Share Tips

A great way to get more customers through Instagram is by sharing tips about your craft. Your audience shouldn't have to pay for the knowledge you have, so why not give them a sneak peek of what they could learn. This could be a fun way to let your followers know what you teach, and how it could help their lives. These are some sneak peeks you could post of your coaching:

- **Habits to learn.** You could post healthy habits that your clients would need to adapt to in order to be submerged in the craft that you're teaching. Give them interesting and useful habits that will improve their quality of life. As a fitness coach, you can let your followers know that they should adapt to healthy lifestyle habits such as drinking sufficient water daily, eating five fruits and veggies every day, and getting enough sleep. These habits let your potential client know what they need to do to get into the craft you're teaching.

- **Outlines.** Many of your followers may not know what you're coaching, and this could prevent them from trying out what you have to offer. Posting an outline of what you teach can give them a good idea of what you do, and from this they can determine whether it's for them or not. Ensure that your outlines are simple and easy to understand, because a complicated outline can put them off your craft. Include fun and exciting things about what you're coaching, because this can entice more clients (no one wants to be coached on something boring).
- **Tips for success.** People choose to be coached at a craft, because they're aiming to learn something new and become good at something that can improve their quality of life. Sharing tips for success can help people being coached by you, and it can give potential clients a sneak peek on what they need to do to be successful. You need to make these tips reasonable and

accessible to everyone, as people like to do things they know they're able to succeed at.

Getting Real

People are often put off by coaches who are loud, flashy, and perfect. This can make you seem fake, which deters potential clients, as no one wants to be scammed. Although it's important to show your good qualities and success as a coach, it's important to be real. You can accomplish this by showing the behind the scenes of your everyday life.

Believe it or not, your followers want to see your everyday life and the struggles that come with it. You can do this by sharing stories and posts that keep people updated on your personal life. For example, as a fitness coach, you can post stories of the cheat meals you've indulged in, and you can express how some days you've broken your consistency, because you've been too tired.

By doing this, you develop a relationship with your followers, because they realize you're just as human as they are. Even as a coach, no one expects you to be a perfect person, so faking it till you make it will only detriment how people view you. However, it's still important to show that you're a motivated, goal-driven coach that they can look up to.

Success Stories

Nothing convinces people more than seeing success stories from other people. Knowing you succeeded at your craft convinces them that you'll be a good coach, but this may not always be enough. Sharing your other clients' success stories can be even more convincing for your followers. Seeing individuals excel from your coaching proves to them that you can get them to their own personal goals.

When you're sharing your clients' success stories, you must be sure to get their consent. Ensure that they're comfortable with having their story shared, and give them the opportunity to write something, or record a video, that best describes their journey. Whenever you're making progress with a client, you should ask them if they're willing to be shown off on your Instagram.

If you haven't had any clients and you're struggling to form a client base, then you must focus on sharing your own success story. You can get people in your life who've witnessed your journey to advocate your growth and success. You could even share success stories of other people who have succeeded in the craft that you're coaching. This can inspire your followers, and make them realize that they can be the next-best success story.

The 4 C's for Coaching

Building an audience on Instagram as a coach is crucial, as it can make or break your success. If you don't build that audience, you may not get enough clients, and this can make it impossible to grow as a coach. Having a big platform with good engagement helps you to reach out to the necessary people who will become your potential clients. You can use the four C's for coaching that can help you to succeed on Instagram.

- **Curiosity.** In order to entice your followers and create a larger client base, you need to pique their curiosity. If someone is curious about your services, it means that they're considering purchasing them. You can inspire curiosity by sharing sneak peeks and revealing the benefits your craft has had on your life.
- **Conversation.** Your Instagram account should not be a one-way act. It's important that your followers know that the communication goes both ways! Encourage your followers to comment, ask questions, and direct message you, as this will also help them to find out more about your coaching and what you have to offer.
- **Community.** Creating a community by using your platform is a great way to build a client base. Making your followers feel like they can be a part of an inclusive community is exciting

for them, and this can influence them to join your coaching.

- **Conversion.** The goal isn't to create quick conversions turning your followers into clients, but rather focus on the relationship you have with your followers. By nurturing these relationships, your followers are more likely to trust you, and this will convert many of your followers into loyal clients.

Utilizing Instagram Strategies

If you're still struggling to create a relationship with your followers, and you notice a lack of conversion rates from followers to clients, then you should consider using Instagram strategies. These strategies could be just what you need to get the client base you want. Some of the Instagram strategies you may want to consider include (Zimmerman, 2018):

- **Improving your bio.** Your bio on Instagram is one of the most important marketing devices. It's like a billboard where you can explain yourself and goals as a coach. Your bio must be brief, but filled with tons of useful information. It should also be written in a fun and friendly tone that shows your followers your personality, but it must also be formal to show potential clients your professional side. The first thing people see when they visit your account will be your bio, so make sure it's great!

- **Question time.** Questions can do so much for your business as a coach, as your clients make your business. Having question time once a week can be a great way to engage with your followers' concerns, interests, and questions. Instagram has a great feature that allows you to put a question box up on your story. Once your followers press the box, they're able to type in any question, suggestion, or concern that may be on their mind. You can then reply to this in their direct messages, or you can share it publicly for everyone to see. You should think of ways to make it fun and interesting, so that it can draw traffic from your audience.
- **Client of the month.** Having competitions with your clients is a good incentive to keep them going. We all love recognition when we're doing well at something, as this can motivate us to keep going. You can monitor all of your clients' progress and dedicate a post on your Instagram for them once a month. You could even give them a discount for how much they need to pay that month. Your followers who aren't clients yet will see that you reward hard workers, and you recognize when people are excelling. This can convince people to join you,

because they want to be a part of a rewarding community.

When you're coming up with the perfect Instagram strategies you should use, you should do your research. Look at all the options out there and decide what will work best for your audience. You could also purchase an Instagram strategy guide that gives you the steps and tricks to building that platform for success!

Consulting Through Social Media

Consulting is a bit different than coaching, because people will only purchase your services when they need them. It's rare for someone to be convinced to use a consultant when they don't have a problem that needs to be fixed. This is why utilizing social media can be really valuable for your consulting business. You need to show people and businesses out there that you're a more affordable way to solve their problems.

How to Get Started With Consulting on Social Media

Anyone can get started as a consultant, as all you need is skill or expertise that you can offer to people. You need to be willing to utilize your problem-solving skills and communicate effectively with your clients. If you're thinking big, you can turn your consulting into a business instead of merely becoming a freelance consultant. This is a step-by-step guide that will help you to establish a consulting business through the use of social media (Aland, 2018).

Step 1: Acquire the Skills You'll Need

If you don't have skills or expertise you can use to become a consultant, you can always learn something new! You must start off by determining what field you'd like to consult in. Maybe you've always wanted to explore your passion in marketing and help businesses meet their marketing potential.

Once you've determined what field you want to consult in, you need to consider what skills and expertise you need in order to be a good consultant. You'll need to know about marketing strategies, and how to apply them to businesses. You must also determine how you'll acquire these skills. When you're getting into consulting, remember that clients will expect the following from you, in order for them to trust you have the skills they need:

- **Qualifications.** Qualifications are valuable proof that show your potential clients that you know what you're talking about. Getting qualifications may sound daunting, because you don't have the time to study for a three-year degree. However, there are courses that you can take that will provide you with the qualifications and skills you need to become a successful consultant. Paying for online courses through platforms like Udemy is a good investment for your consultant career. These courses are short, efficient, and provide you with credentials.

- **Social media presence.** No one wants to use a consulting business that has a dead social media presence, with few to no posts and no following. This can make you look unreliable and fake. It's important to be active on social media, and showcase all of your skills on your platform. Post content that shows people what you do and how good you are at it.
- **Experience.** Experience can add a lot of credentials to your name as a consultant. Clients like to know that you've successfully provided your services to other clients before you, as this makes you seem more reliable. The experience doesn't need to be grand or expansive, as long as you've used your skills professionally before. You could use freelancing platforms like Upwork, to get some independent jobs that can act as experience.

Step 2: Use Local Businesses as Your First Clients

The goal for your consulting business is to get the opportunity to consult for a variety of different businesses, big and small. However, you may see more results by starting small. Local businesses that are smaller may be more open to using your new consulting services, especially if you lack the experience.

As a consultant, your portfolio is your sales pitch, and without one, you may not get the number of clients you aimed for. You should start by approaching these local businesses and asking if your services could be useful to them. You could make your first job free or discounted, and then charge for the next jobs if they want to hire you again. The more local businesses you get, the more your portfolio will grow, and from this you can expand your business to work with bigger clients.

Step 3: Creating Your Rates

Your rates are how much you want to get paid per hour for your services. You make your rate as high or low as you want, but it's important to be realistic. You want to make appealing rates that clients are drawn to. Companies choose to use freelance consultants, because they can be more affordable than businesses, so it's important to consider this when you're making your rates.

Your rates will vary as a consultant, depending on your experience, portfolio, and skill set. If you've got a long list of jobs behind your name, this can justify you increasing your rates. However, it's important to keep your rates competitive, so that you stand out from your competition. The average online consultant's rate ranges between $20–$40. It's important to research the rates of consultants in your field, so that you know what you're competing with.

Step 4: Sticking to Your Niche and Getting Into Freelancing

You know what field you're consulting in, but have you chosen a niche that narrows down your services even more? Sticking to a niche can help you to appeal to more clients, because it makes you seem more skillful and more of an expert. You may be interested in a field, but there are some elements of it that you're not necessarily good at. Choosing a niche that you're great at can ensure that your services are always efficient and professional.

When you're getting into consulting as an individual, freelancing can be your best friend. Using platforms like Fiverr and Upwork will provide you with the clients looking for freelancers to get the job done. Start off with your niche on these sites, and as you grow confidence in your consulting ability, then you can expand your niches and what you deal with. Freelance jobs will also add to your portfolio as experience, and you'll be able to seem more professional as a consultant business.

Step 5: Expand Your Business

After you've paved your way as a freelance, independent consultant, it's time to consider whether you'd like to expand into a business. Turning your consulting into a business will allow you to do less of the hands-on work, while increasing your revenue. When you expand your business, one of the first things you should do is hire help.

Expanding your business means that you need helping hands that will work under you, so that you can get more jobs done at the same time. When you're going through the hiring process, it's important to seek hard workers who will meet your consulting standards. You want all of your clients to experience high-quality service from your business, despite who's working with them.

Step 6: Get External Funding

Expanding your business can be an expensive task, because you have employees to hire and pay. You also need to spend money on marketing and advertising your business, but if you're savvy with your social media, you can find ways to promote yourself for free! You may even want to pay for business courses that will teach you how to run your consultant business successfully.

With all these added expenses as you grow your business, you may want funding that'll help you to become the best consultant business possible. There are many different loans and packages that have low interest rates for start-up businesses. This can help your consultant business to get on its feet.

Tips for Success

Following these steps will help you on your way to becoming a successful consultant, but like any field, it is competitive. Depending on your niche, you'll find a wide range of other businesses and freelancers offering the same services as you. This is why it's important to find ways to stand out from your competition. These are some tips that will help you to embark on a successful consulting career (Victor, 2021):

- **Network.** The best way to get your consulting business out there is by networking with people. This is a free way to promote your business, but it takes time to make the necessary connections you need. It's valuable to start networking as soon as you consider getting into consulting. Let people know in your field that you're offering your services at an affordable hourly rate. You should also encourage your clients to let people know about your affordable and efficient services once you're done working with them. You could get a large inflow of clients through word of mouth.

- **Invest in your Instagram account.** As a business owner, social media is everything. You need to ensure that you invest your time, energy, and even money into creating an Instagram account that best advertises your business. You can pay experienced freelancers to revamp your Instagram account by making it look fresh and inviting. Your money could also go into paying for promotions on Instagram, so that your account can reach a target audience that may be in need of your help.
- **Plan ahead.** As established throughout this chapter, there are expenses that you need to prepare for if you're aiming to become a successful consultant. This makes it vital for you to plan ahead, by creating budgets that break down all the expenses you plan to pay for. This will help you to save up before you get started. Planning your consulting career beforehand can help to prepare you for success.

Chapter 8:

Digital Courses

If you're interested in teaching people, but you don't want to have to work one on one with clients, then you may want to consider creating online courses. Digital courses are popular ways to learn for people who want to learn independently without a physical teacher. This may be a useful way for you to make money online.

Why You Should Make and Sell Online Courses

The great thing about making and selling online courses is that you don't need to be overly skilled and experienced; all you need is an interest in teaching people something useful. However, it does help to have some expertise when you're making your own course from scratch. Selling online courses can be a lucrative way to make a passive income once you get the hang of it (*Who should create an online course and what are the benefits?*, n.d.).

Is Selling Online Courses for You?

As with all these forms of digital creation, it is accessible for everyone, but it may not suit your personal qualities and your lifestyle. You should take the time to consider whether it's for you, because there are a few difficulties that come with creating and selling online courses. These are some things you should ask yourself when you're considering whether creating and selling courses is for you.

- *Do you have a skill or talent to teach?* If you're planning on selling online courses that you're creating, you need to consider whether you have something to teach that people would be interested in. For example, you could have accounting skills to share, or you may be a talented cook with tips and secrets to share. Whatever it is, make sure it's something that could be taught successfully through an online course.
- *Are you interested in starting an online business?* Selling courses online is its own business, especially when you start your own brand out of it. When you approach this side hustle as a business, you're able to make more money from it. This is why it's important for you to consider whether starting an online business is something you'd really want to do. Does the thought of becoming a digital entrepreneur

excite you? If so, then this may be the perfect side hustle for you!

- *Will selling courses benefit my other online business?* If you're on your way toward becoming a digital creator by utilizing some of the digital assets mentioned in this book, then you need to consider whether selling courses online will benefit these online businesses. For example, if you've started a blog about healthy eating, selling an online course about dieting and fitness will benefit your businesses. Your different online businesses can help to promote each other.

Benefits and Perks of Creating and Selling Online Courses

If you answered yes to any of the questions above, you should consider selling online courses. What may convince you is the long list of benefits and perks you gain from starting this online business. We all want to start a side hustle that will only benefit us in the long run and get us a step closer to experiencing financial freedom. These are just a few perks and benefits you can expect from starting your online course business:

- **Achievable source of passive income.** With many avenues of passive income, they still require a fair amount of work and effort, even when your business is up and running. However, selling online courses can become a passive income that doesn't require endless amounts of your time and energy. The most work goes into creating the course, and once it's done, you must market it effectively, and watch the passive income roll in.

- **A wide audience you can reach from anywhere.** The benefit of online courses is that you can reach a wide audience from across the world. This can lead to a larger opportunity for sales. Due to this all being digitized, you also have the benefit of being able to work anywhere in the world. You can work in Italy, with your clientele being in America. This gives you no excuse to choose not to venture into selling digital courses.
- **Limited costs.** As opposed to the physical educational industry, you don't have to worry about the costs for resources. You don't have to pay for paper, textbooks, or manufacturing costs. This can make it a profitable venture, as there aren't endless costs that go into making these courses. Not having to purchase additional materials and resources also helps you live clutter-free, and not to mention it's environmentally friendly!

- **Work on something you're passionate about.** When creating a digital course, you're able to make it on something that interests you. For example, you may be passionate about performing, so you can create a digital course that teaches people the basics of being an entertaining performer. Working on something you're passionate about can make you excited to get up every day to work. You'll also produce courses that you're truly proud of, and this will motivate you to promote it to the best of your ability.
- **Benefits your customers**. Not only will selling online courses benefit you, but it will also benefit your customers. The traditional way of getting an education can be costly, time consuming, and stressful. By doing courses, you're able to learn valuable information without the pressure of being in a school environment. Digital courses are also a lot more affordable and time efficient, helping students to save their money. Your customer may be a student at school, but they can use your digital course to gain more knowledge. These courses can also be fun and informative, helping your customers learn important things about life, hobbies, and activities that can improve their quality of life.

Challenges You May Face

Just as there are positives for this business, there are also some challenges that you may face. It's important to know whether these challenges are something you can tackle. If you feel as though it's something you're prepared to work through, then it's time to get started with your own digital course business! These are some challenges you may face in this industry:

- **Time needed to start up.** Although you're not faced with a lot of start-up costs when selling digital courses, you're faced with the problem of time. It takes a lot of time to make your own digital course that is of good quality. You need to invest this time into making your courses as professional as possible, so that you can grow the reputation of providing great courses. After spending time on creating your digital course, you have to spend time on marketing, and after that, you'll start earning a passive income that doesn't need to be monitored constantly.
- **Investing in marketing.** Marketing is such an important part of selling digital courses. Without the proper marketing you may not receive the results you desire. This is why it's so vital for you to invest your time and money into it. You should research what forms of marketing would be best for your digital course, and from this you can take the necessary actions

to implement these marketing strategies. Unfortunately, you may need to pay a substantial amount of money on your marketing, depending on how professional you want it to be.

- **Taking a course.** It can be difficult to create a digital course that is easy to use, effective, and serves its purpose as a learning tool. It's fundamental to make a course that ticks all these boxes, so you may need to invest in taking an online course for yourself. There are online courses that teach you how to make digital courses, or you could take any digital course that interests you, and you can observe how it's done. Seeing a digital course first-hand helps you to realize what's expected from you when you're creating your own.

How to Create and Sell Online Courses

Creating a digital course is relatively simple, as anyone can do it, but it does take some time and effort. You need to be up for the process and willing to do your research, put in the hours and care about the course you're creating. If you're passionate about the topic your digital course is on, you're going to be motivated to work hard on it. These following steps will give you a good idea of how you can go about creating and selling online courses (Sumrak, 2021).

Step 1: Find the Perfect Opportunity

As with any business, it's important to find the right opportunity that'll get you started. This opportunity will be the foundation of your business, and from this you'll be able to provide what consumers want. When you're trying to identify the perfect opportunity, you need to go through this process:

- **Identifying a problem in the market.** The most successful businesses are created from problems that have been identified in the market. These problems are the demand for products that aren't being provided to consumers. For example, actors and actresses may be needing a course that develops their acting skills, but there aren't any affordable courses that they could take. They may not even have access to physical classes because of where they live.
- **Defining this problem.** To define the problem in the market, you need to determine how this issue is a problem and why it needs to be solved. What is the detriment of this issue being unsolved in the market? For actors and

actresses who need acting courses, they may see a downfall in their performance. They don't have the resources that can help them to improve and become the best actors possible. This can negatively impact their chances of making it in this competitive industry.

- **Providing your solution.** Your solution will be your digital course that you plan on providing. Consider how you can provide a course that solves the problem in the market, this will be your opportunity to get started. If you produce a digital course that provides acting tips and methods that actors and actresses can use to improve their craft. It's important to find ways to adapt your digital course to what you're aiming to provide. For example, you can add a video element to your course to help your students to practice these acting tips and techniques.

In order to find this perfect opportunity, you need to take into consideration what you have interest in. There's no point in working on a digital course just because it's in demand, but you don't have any experience or knowledge in this field. Choose a topic or craft that suits you and your skills, and from there you can determine whether there's an opportunity for it to succeed. You can identify these opportunities by doing the following:

- **Use a survey.** A survey can give you a lot of useful insight to what consumers want. If you already own a digital business with an email list, you can send out surveys to everyone on that list, and see what digital courses they'd like to see you. You can also ask whether they'd support you if you launched a digital course business on the side. If you don't have an email list, you could reach out to social media followers, or you could ask your friends and family to share it with people they know.
- **Visit platforms like Quora.** Quora is a public platform where people are able to share their opinions. Other forums you could visit include Reddit and Stack Overflow. You can upload your digital course idea and receive useful feedback that informs you on the demand of your product, or you could check for any digital courses that have been provided around the same topic. You can review all the questions and comments under these posts, as this will give you an idea of what people are thinking about digital products like yours.
- **Reach out to your followers on social media.** It's a good idea to start a social media account for your digital course business. This can help you to build a following of people who will let

you know whether your digital course is something they'd invest in or not.

Step 2: Validating the Idea

Once you've determined the perfect opportunity to use, you need to work on validating this idea. Is it a lucrative idea that'll get you the number of customers that you're aiming for? This can be determined by doing more in-depth research on your course. This will prevent you from spending time and energy on a digital course that doesn't do well.

You should do the Smoke Test to determine whether your digital course will be a hit or not. Doing a smoke test is when you launch your product without it being done yet. You set up a landing page and see if you get traffic and sales on your page. You'll be able to see how many people clicked on your page, and of those people how many people clicked to purchase your product.

This procedure will take a couple days to get valuable results, and from this, you'll determine whether releasing your digital course is worth the time and effort or not. You should establish a goal of click-through rates you want, that will make you comfortable with producing your online course. If you don't reach this goal, consider developing your digital course idea to make it more relative and in demand.

Step 3: Create an Outline for Your Course

After establishing what your digital course is going to teach, it's time for you to create an outline for it. This is the most challenging part of the process, because it's where you make all the important decisions that will make your course either professional and successful, or below consumer's standards. This is why it's vital to invest your time in it, by paying attention to details.

We live in an age where information is so accessible for anyone, all you need is the Internet. In order to provide a course to your customers that's actually useful to them, you need to provide them with a learning experience that improves their knowledge and skills. If you don't provide them with their experience, then they might as well have searched for this information online for free.

This is how you can create content that will transform people's lives, even though it may not be fresh information:

- **Utilizing existing content.** It's almost impossible to come up with completely fresh content that hasn't been produced before. There's so much content out there, so why not use what already exists? This can make it easier for you to find enough content for your educational courses. When you use existing content, it's important to rewrite the content to ensure that you're avoiding plagiarism. You also want to write your content in a way that best suits the tone and style of your course.

- **Guide your students.** People aren't paying for your course to see fresh, never-before-seen content; instead, they're paying for something that will guide them toward success. They want an experience that will help them to learn, grow, and excel. This makes it crucial for you to make an outline that will guide people to their success. Think about what it is that motivates learning, and helps people to get the results they desire. In order to establish this, you should do efficient research. Ensure that your course has everything someone needs to learn to perfect the craft you're teaching.
- **Focusing on the course structure.** The structure of your course affects how easy or difficult it is to follow for your students. You want it to be set out in a way that makes it easy for customers to learn from. Make sure that it's well planned out with the modules set out logically. You could start off with easier modules, and get progressively difficult, so that the student can adjust to the difficulty level. You should also group all the information that is similar and related into the necessary

modules, as this will make it logical and fluid for your customers.

Step 4: Making the Online Course

Once you have a well-structured, planned outline, it's time to implement this outline by turning it into a digital course. You need to determine how you'd like to deliver your content, because this can impact the effectiveness of your course. Consider what you're trying to teach and think about what form of deliverance is most appropriate. You could deliver it in any of these ways:

- **Written.** Written content in courses is a very popular way of doing things. You could type out all of the content your students will need to learn, and at the end of each module, you can give the student the opportunity to write answers to questions. You could add fun quizzes and questions that promote interaction.
- **Audio.** You could upload your content as audio files. These can be recorded by you or with someone you know that sounds professional. It's important that the audio clips are all audible, and the person speaking pronounces their words properly. You could even hire voice talent who will record the audios for you. These audios can serve as lectures for students, and

they can listen to them while getting other things done.

- **Video.** Many students learn best through the use of videos. This is because they need to be visually stimulated while they're absorbing new content. You may find that videos do better, and help your customers excel with flying colors. You can work on these videos yourself, or you could hire a freelancer to help you. If you don't have experience with making video content, it's advisable for you to get professional help, so that it doesn't look like a school project that your customers are unsatisfied with.
- **Multimedia.** The best courses have a variety of different forms of content. This is more helpful for students, because it helps them to stay interested, and it's inclusive for the different types of learning styles people require. You can have a written description of your module, followed by a lecture that's in the form of an audio, and you can then end off with a fun and engaging video that summarizes the content of that module. You could even add a quiz at the end that the students can answer.

Step 5: Forming a Test Group

Before you launch your course to the public, it's crucial to go through a testing phase. This phase can be accomplished by forming a test group that will provide you with feedback. In order to make a test group that will be useful to your business, you need to get people who you don't know, so that their feedback is honest and objective.

You should also pay your test group a small incentive for taking part in your project, because they could be of use to you further down the line. They could be ambassadors who advertise your digital courses, or they could be success stories that you use to promote your business. People love to see a product work for someone else, before they consider purchasing it for themselves.

Your support group will take your course, and once they're done, they'll provide you with feedback. This feedback is fundamental for your business's success. You must take all the negative feedback you're given and work on improving it. This test run can help you to neaten up and perfect your course, so that paying customers are satisfied with their purchase.

Step 6: Launch Your Digital Course

Once you've successfully accomplished the above steps, it's time for you to finally launch your digital course!

This is arguably the easiest, yet most important, step. When you launch your course, you need to consider how you're going to promote it to your potential customers. This is what you can do to ensure that your digital course reaches the necessary people:

- **Blogs.** If you own a blog, you can announce your release on your own blog. You could create a blog post that's centered around the content of your digital course. For example, if you're launching a course that teaches people the basics of cooking, you can write a blog article about the importance of knowing how to cook.
- **Social media.** Social media can be the best way to promote your course, especially if you've created an account for your business. However, if you don't have your own platform, or your platform doesn't reach a wide audience, you could use an influencer. This is just one form of affiliate marketing that you can use to promote your course. Ensure that you use an influencer who is appropriate for the course you're promoting.
- **Digital advertising.** You can't go wrong with digital advertising, because it can be a simple and effective way to reach your target audience. Technology is so advanced that it can target your adverts to people who actually need them.

The more money you put into this type of digital advertising, the more results you could receive.

Platforms You Can Use

If you want to try out an online course for yourself, so that you can see your competition, and realize what's expected from you as a digital course producer, then you can try out the courses from some of the best platforms including Udemy, Shaw Academy, Skillshare, LinkedIn Learning, Treehouse, and Coursera (Enfroy, n.d.).

Conclusion

You no longer have to settle for that 9–5 job that's wasting your potential, and dampening your dream of financial freedom. It's time to take advantage of the countless resources at your disposal online. Be your own boss by investing in digital assets, and providing useful resources to markets in demand.

Luckily for you, you don't need to have lots of money in the bank or apply for huge loans in order to become the entrepreneur you've always aspired to be! You can start your own business through the use of your cellphone or laptop with very low start-up costs. You can even choose to start multiple digital businesses and create an online empire. There's an endless number of resources available to you, so all you need to do is work hard toward achieving the success you desire.

Aim for Success

"Don't be afraid to give up the good to go for the great."

—John D. Rockefeller

You could choose to use any of these ways of making money online as a side hustle, or you could devote your career to becoming a digital creator. It's important to start small to see if creating digital assets is for you, and if you're happy with the results, take that leap of faith in yourself!

Unfortunately, with a digital career, your income won't be stable and guaranteed, but if you work hard enough, you can make a good living that you're proud of. It's not about fast and easy success, it's about making sacrifices, putting in the time and effort, and getting the result of passive income in return. Don't be disheartened if you don't see the results you want at first, because anything worthwhile takes time.

The luxury of being a digital creator is that you work on your own terms. You establish your working hours, and you can decide where in the world you want to work from. There's such a wide variety of digital assets to choose from, so you have the opportunity to invest your time and energy not something you're passionate about. At the end of the day, it's your passion and discipline that will get you the long-term results you'll be happy with.

References

Abenzour, A. (2019, June 27). How to build great website templates. Blog.duda.co. https://blog.duda.co/how-to-build-great-website-templates

Aland, M. (2018, January 10). 6 Steps to start a social media consulting business ⋆ Egg Marketing. Egg Marketing. https://www.eggmarketingpr.com/blog/6-steps-to-start-a-social-media-consulting-business/

Connolly, B. (2021, June 14). What is Amazon Merch? 2021 Best practices & tips for success. Jungle Scout. https://www.junglescout.com/blog/merch-by-amazon/

Dean, G. (2019, August 29). 15 Ways to make money on fiverr. Forbes. https://www.forbes.com/sites/gingerdean/2016/08/29/15-ways-to-make-money-on-fiverr/?sh=4937d3995cda

Duermyer, R. (2020, January 17). Is affiliate marketing really profitable? The Balance Small Business. https://www.thebalancesmb.com/can-you-really-make-money-with-affiliate-marketing-1794168

Dunne, C. (2017, April 13). How does Merch By Amazon work? (Expert tips for 2021). RepricerExpress. https://www.repricerexpress.com/how-merch-by-amazon-work/

Enfroy, A. (n.d.). 13+ Best online course platforms (Ultimate Guide for 2020). Adam Enfroy. https://www.adamenfroy.com/best-online-course-platforms

Hartgers, A. (2019, December 9). Self-Publishing: A beginner's guide to what is self-publishing in 2020. SelfPublishing.com : The #1 Resource for Self-Publishing a Book. https://selfpublishing.com/self-publishing/

Haysom, S. (2020, February 24). These self-published authors are actually making a living. Here's how. Mashable. https://mashable.com/article/self-published-authors-making-a-living

Helms, C. (2018, November 30). What is Upwork? Medium. https://christianmhelms.medium.com/what-is-upwork-4e3a5d933f1

https://www.facebook.com/adamenfroydotcom. (2018, July 27). Affiliate marketing in 2020: What it is + How beginners can start. The BigCommerce Blog. https://www.bigcommerce.com/blog/affiliate-marketing/#benefits-to-joining-the-bigcommerce-affiliate-program

Mama & Money. (2020, August 12). What to sell on etsy to make passive income. Mama & Money. https://mamaandmoney.com/what-to-sell-on-etsy/

Morgan, S. (2014, June 26). 5 ways to share your coaching skills on Instagram. Coaching Blueprint - Marketing for Life Coaches. https://www.coachingblueprint.com/instagram-for-life-coaches/

Noorani, S. (2019, February 19). How to make money by selling PLR eBooks. Let's Earn Money Online. https://www.letsearnmoneyonline.in/make-money-selling-plr-ebooks/

Patel, N. (2015). Affiliate marketing made simple: A step-by-step guide. Neil Patel. https://neilpatel.com/what-is-affiliate-marketing/

Pilon, A. (2018, March 8). How one female entrepreneur built a business on merch by amazon. Small Business Trends. https://smallbiztrends.com/2018/03/business-on-merch-by-amazon.html

Schulz, S. (2021, January 25). My honest review of AND.CO [Fiverr's app for freelancers]. Don't Do It Yourself. https://ddiy.co/and-co-review/

Sheldon, S. (n.d.). What is PLR content? | The value of PLR. Contentsparks.com. Retrieved August 27, 2021, from https://contentsparks.com/953/what-is-plr-content/

Sumrak, J. (2021, June 14). How to create an awesome online course in 2021. Foundr. https://foundr.com/articles/building-a-business/online-courses/create-online-course

Thackston, K. (2020, April 22). 7 Benefits of PLR content: Create custom blog posts easier & faster. Marketing Words Blog. https://www.marketingwords.com/blog/plr-content/

The ultimate list of digital products to sell on etsy. (2019, October 11). Ebb and Flow Creative Co. https://ebbandflowcc.com/digital-products-to-sell-on-etsy/

Truex, L. (2018, June 20). How to use private label rights to grow your home business. The Balance Small Business. https://www.thebalancesmb.com/private-label-rights-for-home-business-1794347

Victor, M. (2021, June 25). How to start a social media consulting business in 2021. The Penny Matters. https://thepennymatters.com/social-media-consulting-business/

WayToChanges. (2020, March 16). How to use PLR products: The ultimate guide. Way to Changes. https://www.waytochanges.com/how-to-use-plr-products/

Who should create an online course and what are the benefits? (n.d.). Podia. Retrieved August 28, 2021, from https://www.podia.com/how-to-create-sell-profitable-online-course/benefits

Zimmerman, S. (2018, November 16). Get coaching clients (INSTAGRAM TIPS TO BOOK MORE SALES). SBZ Enterprise. https://suebzimmerman.com/get-coaching-clients-instagram-tips-to-book-more-sales/

Made in United States
North Haven, CT
05 November 2026

21790517R00104